Iosias Jody (Ed.)

Klingnauer Stausee

Iosias Jody (Ed.)

Klingnauer Stausee

Böttstein, Zurzach District, Baden, Switzerland, La Tène culture

Cred Press

Imprint

Publisher:
Cred Press is a trademark of
International Book Market Service Ltd., 17 Rue Meldrum, Beau Bassin, 1713-01 Mauritius
Email: info@bookmarketservice.com
Website: www.bookmarketservice.com

Published in 2011

Printed in: U.S.A., U.K., Germany. This book was not produced in Mauritius.

ISBN: 978-613-7-31399-2

Contents

Articles

References

Klingnauer Stausee

Klingnauer Stausee	

Location	Aargau
Coordinates	47°35′N 8°14′E
Lake type	reservoir
Primary inflows	Aar
Primary outflows	Aar
Basin countries	Switzerland
Max. length	3 km
Max. width	500 m
Surface area	1.16 km²
Max. depth	8.5 m
Residence time	0.13 days
Surface elevation	318 m
Islands	1
Settlements	Döttingen, Klingnau

The **Klingnauer Stausee** (or "Klingnau reservoir") is a reservoir near Böttstein, canton of Aargau, Switzerland, at 47°35′N 8°14′E. The lake with a surface of 1.16 km² was formed at the construction of a power plant on the Aar river in the 1930s.

External links

- http://www.klingnauerstausee.ch

Böttstein

Böttstein	

Country Switzerland **Canton** Aargau **District** Zurzach 47°34′N 8°14′E	
Population	3691 (Dec 2010)[1]
- Density	498 /km² (1290 /sq mi)
Area	7.41 km² (2.86 sq mi)
Elevation	356 m (1168 ft)
Postal code	5315 (Böttstein) 5314 (Kleindöttingen)
SFOS number	4303
Mayor	Agnes Canonica
Surrounded by	Döttingen, Klingnau, Leuggern, Mandach, Villigen, Würenlingen
Website	www.boettstein.ch [2] SFSO statistics [3]
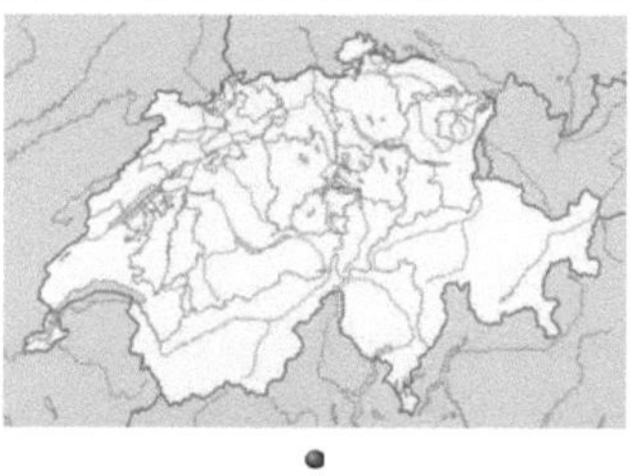 **Böttstein**	

View map of Böttstein

Böttstein is a municipality in the district of Zurzach in the canton of Aargau in Switzerland.

History

Some scattered La Tène culture items have been discovered near Böttstein. The modern village of Böttstein is first mentioned in 1087 as *Botistein*.[4] During the 11th to 13th Centuries it was the seat of the lords of Böttstein, who ruled over the village. The other, neighboring villages were under the lower court of the Knights Hospitaller order house at Leuggern. After the conquest of the Aargau in 1415, the *Herrschaft* was held by the district of Leuggern in the County of Baden. The ownership of the administrative rights of the village wend through about ten changes until 1606 when it was bought by the von Roll family of Uri. Between 1615-17, they replaced the original castle with a late-Gothic - early Baroque mansion with chapel. Through marriage, the property came in 1674 to the Schmid family of Bellikon, who ruled over the village until 1798 and occupied the castle until 1893. The chapel of St. Anthony was given a wide range of ornamentation in the manner of Italian baroque churches with excellent stucco work, all done by northern Italian masters. In 1816 the municipality was formed by dividing the large municipality of Leuggern.[4]

Böttstein and the other villages were part of the parish of Leuggern.

Until the late 19th Century the major economic activities were farming, viticulture and handicrafts. During the late 19th Century, the timber industry grew along the Aare river, which provided many new jobs. The new jobs along the river allowed the small village of Kleindöttingen to expande rapidly. Despite its own industry (1970-94 particle board factory, furniture factories) and numerous commercial enterprises the municipality is primarily a commuter town.[4]

Geography

Böttstein has an area, as of 2009, of 7.41 square kilometers (2.86 sq mi). Of this area, 2.52 km^2 (0.97 sq mi) or 34.0% is used for agricultural purposes, while 2.6 km^2 (1.0 sq mi) or 35.1% is forested. Of the rest of the land, 1.54 km^2 (0.59 sq mi) or 20.8% is settled (buildings or roads), 0.68 km^2 (0.26 sq mi) or 9.2% is either rivers or lakes and 0.11 km^2 (27 acres) or 1.5% is unproductive land.[5]

Of the built up area, industrial buildings made up 3.6% of the total area while housing and buildings made up 8.9% and transportation infrastructure made up 4.0%. Power and water infrastructure as well as other special developed areas made up 3.1% of the area while parks, green belts and sports fields made up 1.1%. Out of the forested land, 32.8% of the total land area is heavily forested and 2.3% is covered with orchards or small clusters of trees. Of the agricultural land, 22.0% is used for growing crops and 9.9% is pastures, while 2.2% is used for orchards or vine crops. Of the water in the municipality, 3.6% is in lakes and 5.5% is in rivers and streams.[5]

The municipality is located in the Zurzach district, on the left bank of the Aare river a few kilometers from its confluence. It consists of the villages of Böttstein and Kleindöttingen and the hamlets of Eien and Burlen.

Coat of arms

The blazon of the municipal coat of arms is *Or Frete Gules.*[6]

Demographics

Böttstein has a population (as of December 2010) of 3691[1] As of June 2009, 39.2% of the population are foreign nationals.[7] Over the last 10 years (1997–2007) the population has changed at a rate of 1.1%. Most of the population (as of 2000) speaks German(80.3%), with Italian being second most common (6.3%) and Albanian being third (4.3%).[8]

The age distribution, as of 2008, in Böttstein is; 368 children or 9.8% of the population are between 0 and 9 years old and 502 teenagers or 13.3% are between 10 and 19. Of the adult population, 500 people or 13.3% of the population are between 20 and 29 years old. 437 people or 11.6% are between 30 and 39, 614 people or 16.3% are between 40 and 49, and 573 people or 15.2% are between 50 and 59. The senior population distribution is 405 people or 10.8% of the population are between 60 and 69 years old, 262 people or 7.0% are between 70 and 79, there are 91 people or 2.4% who are between 80 and 89,and there are 12 people or 0.3% who are 90 and older.[9]

As of 2000 the average number of residents per living room was 0.63 which is more people per room than the cantonal average of 0.57 per room. In this case, a room is defined as space of a housing unit of at least 4 m^2 (43 sq ft) as normal bedrooms, dining rooms, living rooms, kitchens and habitable cellars and attics.[10] About 52.2% of the total households were owner occupied, or in other words did not pay rent (though they may have a mortgage or a rent-to-own agreement).[11]

As of 2000, there were 132 homes with 1 or 2 persons in the household, 762 homes with 3 or 4 persons in the household, and 471 homes with 5 or more persons in the household.[12] As of 2000, there were 1,392 private households (homes and apartments) in the municipality, and an average of 2.6 persons per household.[8] In 2008 there were 433 single family homes (or 27.6% of the total) out of a total of 1,567 homes and apartments.[13] There were a total of 4 empty apartments for a 0.3% vacancy rate.[13] As of 2007, the construction rate of new housing units was 0 new units per 1000 residents.[8]

In the 2007 federal election the most popular party was the SVP which received 49.84% of the vote. The next three most popular parties were the CVP (20.11%), the FDP (10.67%) and the SP (10.47%). In the federal election, a total of 962 votes were cast, and the voter turnout was 48.7%.[14]

The historical population is given in the following table:[4]

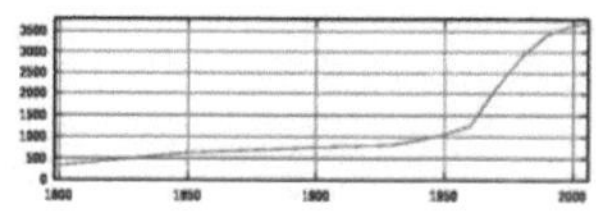

Population

year	population
1799	Böttstein 168 Kleindöttingen 96 Eien 60
1850	together 629
1888	536
1900	760
1950	1,051
2000	3,648

Heritage sites of national significance

Schloss Böttstein with the chapel tower visible

The oil press and sawmill at Schlossweg 2 and the castle chapel are listed as Swiss heritage sites of national significance.[15] The entire village of Böttstein is designated as part of the Inventory of Swiss Heritage Sites.[16]

Economy

As of 2007, Böttstein had an unemployment rate of 2.52%. As of 2005, there were 43 people employed in the primary economic sector and about 15 businesses involved in this sector. 548 people are employed in the secondary sector and there are 36 businesses in this sector. 629 people are employed in the tertiary sector, with 91 businesses in this sector.[8]

In 2000 there were 1,827 workers who lived in the municipality. Of these, 1,402 or about 76.7% of the residents worked outside Böttstein while 722 people commuted into the municipality for work. There were a total of 1,147 jobs (of at least 6 hours per week) in the municipality.[17] Of the working population, 15.4% used public transportation to get to work, and 55.3% used a private car.[8]

Religion

From the 2000 census, 2,087 or 57.2% were Roman Catholic, while 661 or 18.1% belonged to the Swiss Reformed Church. Of the rest of the population, there were 7 individuals (or about 0.19% of the population) who belonged to the Christian Catholic faith.[12]

Education

The entire Swiss population is generally well educated. In Böttstein about 62.2% of the population (between age 25-64) have completed either non-mandatory upper secondary education or additional higher education (either University or a *Fachhochschule*).[8] Of the school age population (in the 2008/2009 school year), there are 261 students attending primary school, there are 202 students attending secondary school in the municipality.[12]

References

[1] Canton of Aargau Statistical Office (http://www.ag.ch/staag/daten/B01/index.html), MS Excel document – *Bevölkerungsbestand per 31.12.2010 nach Gemeinde, Geschlecht u. Nationalität* (German) accessed 11 May 2011

[2] http://www.boettstein.ch

[3] http://www.bfs.admin.ch/bfs/portal/en/index/regionen/regionalportraets/gemeindesuche.html?geographyID=4303&FormEncoding=UTF-8

[4] *Böttstein* in German (http://www.hls-dhs-dss.ch/textes/d/D1831.php), French (http://www.hls-dhs-dss.ch/textes/f/F1831.php) and Italian (http://www.hls-dhs-dss.ch/textes/i/I1831.php) in the online *Historical Dictionary of Switzerland*.

[5] Swiss Federal Statistical Office-Land Use Statistics (http://www.bfs.admin.ch/bfs/portal/de/index/themen/02/03/blank/data/gemeindedaten.html) 2009 data (German) accessed 25 March 2010

[6] Flags of the World.com (http://www.fotw.us/flags/ch-ag212.html) accessed 16-June-2010

[7] Statistical Department of Canton Aargau -Bereich 01 -Bevölkerung (http://www.ag.ch/staag/) (German) accessed 20 January 2010

[8] Swiss Federal Statistical Office (http://www.bfs.admin.ch/bfs/portal/en/index/regionen/02/gemeindesuche.html) accessed 16-June-2010

[9] Statistical Department of Canton Aargau -Bevölkerungsdaten für den Kanton Aargau und die Gemeinden (Archiv) (http://www.ag.ch/staag/) (German) accessed 20 January 2010

[10] Eurostat. "Housing (SA1)" (http://www.bfs.admin.ch/bfs/portal/en/index/international/22/lexi.Document.116365.pdf) (pdf). *Urban Audit Glossary*. 2007. p. 18. . Retrieved 12 February 2010.

[11] Urban Audit Glossary pg 17

[12] Statistical Department of Canton Aargau - Aargauer Zahlen 2009 (http://www.ag.ch/staag/) (German) accessed 20 January 2010

[13] Statistical Department of Canton Aargau (http://www.ag.ch/staag/) (German) accessed 20 January 2010

[14] Swiss Federal Statistical Office, *Nationalratswahlen 2007: Stärke der Parteien und Wahlbeteiligung, nach Gemeinden/Bezirk/Canton* (http://www.bfs.admin.ch/bfs/portal/de/index/themen/17/02/blank/data/04/03.html) (German) accessed 28 May 2010

[15] "Kantonsliste A-Objekte: Aargau" (http://www.bevoelkerungsschutz.admin.ch/internet/bs/de/home/themen/kgs/kgs_inventar/a-objekte.html) (in German). *KGS Inventar* (http://www.bevoelkerungsschutz.admin.ch/internet/bs/de/home/themen/kgs/kgs_inventar.html). Federal Office of Civil Protection. 2009. . Retrieved 28 May 2010.

[16] ISOS site (http://www.isos.ch/en/nationale/index.asp) accessed 16-June-2010

[17] Statistical Department of Canton Aargau-Bereich 11 Verkehr und Nachrichtenwesen (http://www.ag.ch/staag/) (German) accessed 21 January 2010

- Sarah Brian Scherer, Dominik Sauerländer, Andreas Steigmeier: *Das Kirchspiel Leuggern – Geschichte von Böttstein, Full-Reuenthal, Leibstadt und Leuggern,* 2001

External links

- http://www.boettstein.ch Official website (in German)
- Andreas Steigmeier: *Böttstein* in German (http://www.hls-dhs-dss.ch/textes/d/D1831.php), French (http://www.hls-dhs-dss.ch/textes/f/F1831.php) and Italian (http://www.hls-dhs-dss.ch/textes/i/I1831.php) in the online *Historical Dictionary of Switzerland.*

Zurzach District

Zurzach District Bezirk Zurzach	
— District —	

Country	Switzerland
Canton	Aargau
Capital	Bad Zurzach
Area	
- Total	129.99 km^2 (50.2 sq mi)
Population (2010)	
- Total	32238
- Density	248/km^2 (642.3/sq mi)
Time zone	CET (UTC+1)
- Summer (DST)	CEST (UTC+2)
Municipalities	23

Zurzach District is a district in the Swiss Canton of Aargau with the administrative capital of Zurzach. It covers the Studenlands and is located in the northeastern part of the canton. The district capital is Bad Zurzach. The main valleys are the Surb and the Aare. It has a population of 32238 (as of 31 December 2010).[1]

Geography

The Zurzach district has an area, as of 2009, of 129.99 square kilometers (50.19 sq mi). Of this area, 57.08 km^2 (22.04 sq mi) or 43.9% is used for agricultural purposes, while 51.23 km^2 (19.78 sq mi) or 39.4% is forested. Of the rest of the land, 16.62 km^2 (6.42 sq mi) or 12.8% is settled (buildings or roads).[2]

Demographics

The Zurzach district has a population (as of December 2010) of 32238.[1] As of June 2009, 25.7% of the population are foreign nationals.[3]

Economy

In 2000 there were 15,454 workers who lived in the district. Of these, 11,295 or about 73.1% of the residents worked outside the district while 6,024 people commuted into the district for work. There were a total of 10,183 jobs (of at least 6 hours per week) in the district.[4]

Religion

From the 2000 census, 16,378 or 54.3% were Roman Catholic, while 7,440 or 24.7% belonged to the Swiss Reformed Church. Of the rest of the population, there were 65 individuals (or about 0.22% of the population) who belonged to the Christian Catholic faith.[2]

Education

The entire Swiss population is generally well educated. Of the school age population (in the 2008/2009 school year), there are 2,405 students attending primary school, there are 972 students attending secondary school, there are 756 students attending tertiary or university level schooling in the municipality.[2]

Municipalities

Coat of arms	Municipality	Population (31 December 2010)[1]	Area, km²[5]
	Baldingen	291	2.83
	Böbikon	167	2.59
	Böttstein	3691	7.41
	Döttingen	3749	6.92
	Endingen	2009	8.46
	Fisibach	386	5.79
	Full-Reuenthal	805	4.82
	Kaiserstuhl	398	0.32
	Klingnau	3103	6.71
	Koblenz	1592	4.08
	Leibstadt	1321	6.39
	Lengnau	2541	12.67
	Leuggern	2123	13.76

	Mellikon	241	2.71
	Rekingen	962	3.1
	Rietheim	728	3.93
	Rümikon	242	2.9
	Schneisingen	1339	8.25
	Siglistorf	627	5.51
	Tegerfelden	1062	7.11
	Unterendingen	371	3.46
	Wislikofen	329	3.75
Total		32238	129.99

References

[1] Canton of Aargau Statistical Office (http://www.ag.ch/staag/daten/B01/index.html), MS Excel document – *Bevölkerungsbestand per 31.12.2010 nach Gemeinde, Geschlecht u. Nationalität* (German) accessed 11 May 2011

[2] Statistical Department of Canton Aargau - Aargauer Zahlen 2009 (http://www.ag.ch/staag/) (German) accessed 20 January 2010

[3] Statistical Department of Canton Aargau -Bereich 01 -Bevölkerung (http://www.ag.ch/staag/) (German) accessed 20 January 2010

[4] Statistical Department of Canton Aargau-Bereich 11 Verkehr und Nachrichtenwesen (http://www.ag.ch/staag/) (German) accessed 21 January 2010

[5] Swiss Federal Statistical Office-Land Use Statistics (http://www.bfs.admin.ch/bfs/portal/de/index/themen/02/03/blank/data/gemeindedaten.html) 2009 data (German) accessed 25 March 2010

Baden, Switzerland

Baden

Country Switzerland

Canton Aargau

District Baden

47°28′N 8°18′E

Population	18059 (Dec 2010)[1]
- Density	1371 /km^2 (3551 /sq mi)
Area	13.17 km^2 (5.08 sq mi)
Elevation	381 m (1250 ft)
Postal code	5400
SFOS number	4021
Mayor (list)	Stephan Attiger (as of 2008) FDP/PRD
Surrounded by	Birmenstorf, Ennetbaden, Fislisbach, Gebenstorf, Mellingen, Neuenhof, Obersiggenthal, Turgi, Wettingen
Twin towns	Sighisoara (Romania)
Website	www.baden.ch [2] SFSO statistics [3]

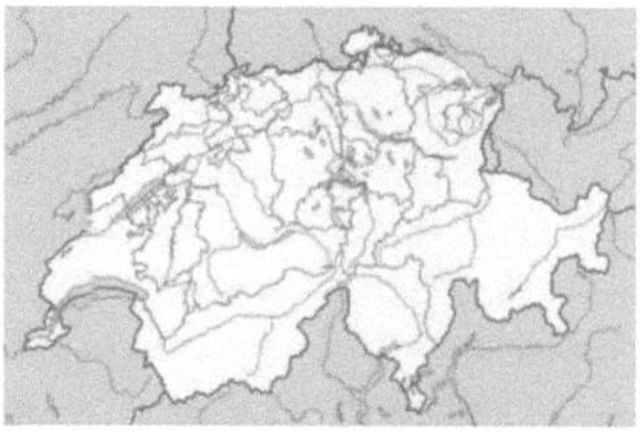

Baden

View map of Baden

Baden is a municipality in the Swiss canton of Aargau, on the west bank of the river Limmat, located in the Limmat Valley (German: *Limmattal*), 25 km (16 mi) northwest of Zürich. It is the seat of the district of Baden. The name refers to the mineral hot springs, which were known at least as far back as the Roman era.

Its population in 2010 was over 18,000.

History

Roman inscription from Aquae Helveticae, now in St. Sebastian Church in Wettingen

Baden in 1642

Baden was known to the Romans as *Aquae Helveticae*. In 1040 the town is first mentioned as Baden.[4]

Aquae Helveticae was first mentioned by Roman historian Tacitus. At that time it was a Roman Vicus to the north of the Baden gorge on the Haselfeld. On the left bank of the Limmat bend there was a pool complex, which was fed by the 47 °C (117 °F) water from a system of springs. The Vicus was founded to support the establishment of the legionary camp at Vindonissa. Both the spa quarter and the residential and commercial district grew to a respectable size in the first half of the first Century AD. The town suffered a setback in 69 AD as members of the XXI Legion burned the wooden buildings of the vicus during the Year of the Four Emperors. The town was rebuilt in stone. When the legion camp at Vindonissa closed in 101 AD, the the town shrank again slightly. Although the prosperity of the local spas diminished, the town survived on trade. The pottery workshop of Reginus and the bronze works of Gemellianus flourished in the second half of the 2nd Century. The multiple Alemanni invasions in the different ideas about the middle of the 3rd Century brought an end to settlement at Haselfeld. However, the construction of a fortification to protect the pools, along with a large number of coins from the 4th century which are stamped with references to the hot springs show that the spa area remained settled in late antiquity and was frequented.[4]

The main axis of the Vicus was the street which came from Vindonissa. On both sides of this road, which ran parallel to the edge of the steep slope, there were residential and commercial buildings. The road was flanked by porticos, behind which were the retail stores. Behind the commercial buildings were the living quarters, adjoining farms, work spaces and outbuildings. In the center of the settlement were some villa-like buildings. Modernly, the remains of three thermal bathing pools have been discovered.[4]

Stein Castle, formerly a stronghold of the Habsburgs, was destroyed in 1415 and again in 1712. In 1415, Baden (along with Aargau) was conquered by the Eight Swiss Confederates, whose bailiff inhabited the other castle, on the right bank of the Limmat, which defends the ancient bridge across the river. As the conquest of the Aargau was the first made by the Confederates, their delegates (or the federal diet) naturally met at Baden, from 1426 to about 1712, to settle matters relating to these subject lands, so that during that period Baden was effectively the capital of Switzerland. The diet sat in the beautifully carved diet hall in the town-hall or *Rathaus*, which can be visited. In 1714 the Treaty of Baden was signed, which put to an end to the war between France and the Holy Roman Empire, and thus completed the treaty of Utrecht (1713). Baden was the capital of the canton of Baden, from 1798 until 1803, when the canton of Aargau was created.

Having attained popularity as a Spa and 'Kur-Ort' (Cure resort) during the 15th and 16th centuries, Baden remained popular and was later frequented by an array of prominent people including Johann Wolfgang Goethe, Friedrich Nietzsche, Thomas Mann, and Hermann Hesse, who was to become a particular devotee, visiting the town annually over a period of almost 30 years.

Geography

Baden's old city nestled in the hills

Baden has an area, as of 2006, of 13.2 km^2 (5.1 sq mi). Of this area, 9.9% is used for agricultural purposes, while 56.9% is forested. Of the rest of the land, 31.6% is settled (buildings or roads) and the remainder (1.6%) is non-productive (rivers or lakes).[5]

Baden is the capital of the district of Baden. The original village is located on the left bank of the Limmat river, with the town sections of Kappelerhof, Allmend, Meierhof and Chrüzliberg as well as the, independent until 1962, village of Dättwil. On the right bank of the river is the village of Ennetbaden (formerly *Kleine Bäder*). The municipalities of Baden and Neuenhof are considering a merger on 1 January 2012 into a new municipality which will be known as Baden.[6]

Demographics

Baden has a population (as of 31 December 2010) of 18059.[1] As of 2008, 26.1% of the population was made up of foreign nationals.[7] Over the last 10 years the population has grown at a rate of 5.8%. Most of the population (as of 2000) speaks German (83.8%), with Italian being second most common (3.3%) and Serbo-Croatian being third (3.0%).[5]

The age distribution, as of 2008, in Baden is; 1,469 children or 8.3% of the population are between 0 and 9 years old and 1,486 teenagers or 8.4% are between 10 and 19. Of the adult population, 3,013 people or 17.1% of the population are between 20 and 29 years old. 3,248 people or 18.4% are between 30 and 39, 2,747 people or 15.6% are between 40 and 49, and 2,151 people or 12.2% are between 50 and 59. The senior population distribution is 1,787 people or 10.1% of the population are between 60 and 69 years old, 1,039 people or 5.9% are between 70 and 79, there are 581 people or 3.3% who are between 80 and 89,and there are 120 people or 0.7% who are 90 and older.[8]

As of 2000, there were 1,377 homes with 1 or 2 persons in the household, 3,960 homes with 3 or 4 persons in the household, and 1,893 homes with 5 or more persons in the household. The average number of people per household was 2.07 individuals.[9] In 2008 there were 1,670 single family homes (or 18.1% of the total) out of a total of 9,217 homes and apartments.[10]

In the 2007 federal election the most popular party was the SP which received 24.7% of the vote. The next three most popular parties were the SVP (21.2%), the FDP (16.4%) and the Green Party (15.6%).[5]

The entire Swiss population is generally well educated. In Baden about 79.5% of the population (between age 25-64) have completed either non-mandatory upper secondary education or additional higher education (either University or a *Fachhochschule*).[5] Of the school age population (in the 2008/2009 school year), there are 995 students attending primary school, there are 377 students attending secondary school, there are 633 students attending tertiary or university level schooling, there are 25 students who are seeking a job after school in the municipality.[9]

The historical population is given in the following table:[4]

year	population
16th - 18th Centuries	1,000 - 1,800
1799	1,202
1850	3,159
1900	6,489
1950	12,127
1960	14,553
2000	16,270

Economy

Baden with the railway station

As of 2007, Baden had an unemployment rate of 2.54%. As of 2005, there were 64 people employed in the primary economic sector and about 16 businesses involved in this sector. 7,745 people are employed in the secondary sector and there are 178 businesses in this sector. 13,933 people are employed in the tertiary sector, with 1,337 businesses in this sector.[5]

As of 2000 there were 9,223 total workers who lived in the municipality. Of these, 5,567 or about 60.4% of the residents worked outside Baden while 15,103 people commuted into the municipality for work. There were a total of 18,759 jobs (of at least 6 hours per week) in the municipality.[11]

In the 19th and 20th century Baden became an industrial town, main seat of the former Brown Boveri Company. Most industrial facilities have moved, but Baden is still the seat of many of the engineering services of ABB and the power service arm of Alstom. The former industrial quarter to the north of the city is now being redeveloped into offices, shopping and leisure facilities.

There is also a Casino in Baden.[12]

Coat of arms

The blazon of the municipal coat of arms is *Argent a Pale Sable and a chief Gules.*[13]

Sights

Baden is chiefly renowned for its hot sulphur springs, which are mentioned by Tacitus (*Histories* i. cap. 7), and are considered to be especially beneficial for cases of gout or rheumatic problems. The main Spa lies a little to the north of the old town beside the river Limmat. Many Roman remains have been found in and around the beautiful gardens of the *Kursaal*. The town, dominated on the west by the ruined castle of Stein, is very picturesque, with steep and narrow streets, a Medieval wooden bridge and one surviving gateway. The old town, the *Tagsatzung* room in the city hall, the 1847 railway station and the building of the *Stiftung Langmatt* are listed as heritage sites of national significance.[14]

In addition to the Roman city, the ruins of castle Stein and the other sites listed above, Baden is home to a number of other Swiss Heritage Sites. The industrial sites include the ABB Schweiz archive along with the former offices of Brown Boveri Company as well as the regional former utilities plant on Haselstrasse 15. There are three designated religious buildings in Baden; the Catholic city church and Sebastians chapel, the Swiss Reformed parish church and the Synagoge on Parkstrasse 17. Perhaps included in the last two groups is the Crematorium and memorial hall on Zürcherstrasse 108. The wooden bridge between Untere Halde and Wettingerstrasse is also included in the list. A number of individual buildings are also included in the inventory. These include; *Bernerhaus* at Weite Gasse 13, Haus *Zum Schwert* on Schwertstrasse or Oelrainstrasse 29, the Hotel Verenahof, the Hotel *Zum wilden Mann*, the spa-theater with a glass foyer at Parkstrasse 20, the Restaurant *Paradies* on Cordulaplatz, Villa Boveri (since 1943 Clubhaus BBC/ABB) and the Villa Langmatt (now an art museum) at Römerstrasse 30.

The village of Baden is designated as part of the Inventory of Swiss Heritage Sites.[15]

Ruins of Stein Castle

City Church of Baden

Swiss Reformed church of Baden

Villa Langmatt, now an art museum

Stadthaus, part of the city hall complex

Tagsatzung from 1531

Wooden bridge over the Limmat

Baden regional power plant, part of the utilities plant

Surrounding area

One mile south of Baden, on the Limmat, is the famous Cistercian monastery of Wettingen (1227-1841), with splendid old painted glass in the cloisters and magnificent early 17th century carved stalls in the choir of the church. Six miles W. of Baden is the small town of Brugg (9,500 inhabitants) in a fine position on the Aare, and close to the remains of the Roman colony of Vindonissa (Windisch), as well as to the monastery (founded 1310) of Königsfelden, formerly the burial-place of the early Habsburgs (the castle of Habsburg is but a short way off), still retaining much fine medieval painted glass.

Transportation

Baden was the destination of the first railway in Switzerland, the *Spanisch Brötli Bahn* transporting the richer people from Zürich to the baths of Baden. Today Baden is a regular stop on the railway lines **Zürich-Basel** and **Zürich-Bern**. Baden is a stop of the S-Bahn Zürich on the line S12 and a terminal station on the line S6.

Train station

The A1 motorway tunnel Baregg is a major junction in the area. It was undergoing construction until 2004 and has been subject to controversy. In 2003, a third tunnel hole was opened to vehicles on the motorway.

Notable people

- Charles Eugene Lancelot Brown (1863–1924), Joint founder of BBC (today ABB) and freeman
- Walter Boveri (1865–1924), Joint founder of BBC (today ABB) and freeman
- Thomas Erastus (1524–1583), Renaissance Physician and Reformed Theologian
- Albert Hofmann (1906–2008); Chemist and discoverer of LSD
- Rainer Huber (*1948), Cantonal councillor of canton Aargau
- Joel Mall (1991-)Goalkeeper for FC Aarau

Football

FC Baden is the local football team. They play their home games at the Esp Stadium in Fislisbach, a short distance from Baden.

Religion

From the 2000 census, 7,059 or 43.4% are Roman Catholic, while 4,636 or 28.5% belonged to the Swiss Reformed Church. Of the rest of the population, there are 31 individuals (or about 0.19% of the population) who belong to the Christian Catholic faith.[9]

See also

- List of mayors of Baden, Switzerland

References

[1] Canton of Aargau Statistical Office (http://www.ag.ch/staag/daten/B01/index.html), MS Excel document – *Bevölkerungsbestand per 31.12.2010 nach Gemeinde, Geschlecht u. Nationalität* (German) accessed 11 May 2011

[2] http://www.baden.ch

[3] http://www.bfs.admin.ch/bfs/portal/en/index/regionen/regionalportraets/gemeindesuche.html?geographyID=4021&FormEncoding=UTF-8

[4] *Baden* in German (http://www.hls-dhs-dss.ch/textes/d/D1633.php), French (http://www.hls-dhs-dss.ch/textes/f/F1633.php) and Italian (http://www.hls-dhs-dss.ch/textes/i/I1633.php) in the online *Historical Dictionary of Switzerland.*

[5] Swiss Federal Statistical Office (http://www.bfs.admin.ch/bfs/portal/en/index/regionen/02/gemeindesuche.html) accessed 28-January-2010

[6] Amtliches Gemeindeverzeichnis der Schweiz (http://www.bfs.admin.ch/bfs/portal/de/index/infothek/nomenklaturen/blank/blank/gem_liste/03.html) published by the Swiss Federal Statistical Office (German) accessed 14 January 2010

[7] Statistical Department of Canton Aargau -Bereich 01 -Bevölkerung (http://www.ag.ch/staag/) (German) accessed 20 January 2010

[8] Statistical Department of Canton Aargau -Bevölkerungsdaten für den Kanton Aargau und die Gemeinden (Archiv) (http://www.ag.ch/staag/) (German) accessed 20 January 2010

[9] Statistical Department of Canton Aargau - Aargauer Zahlen 2009 (http://www.ag.ch/staag/) (German) accessed 20 January 2010

[10] Statistical Department of Canton Aargau (http://www.ag.ch/staag/) (German) accessed 20 January 2010

[11] Statistical Department of Canton Aargau-Bereich 11 Verkehr und Nachrichtenwesen (http://www.ag.ch/staag/) (German) accessed 21 January 2010

[12] Grand Casino Baden website (http://www.grandcasinobaden.ch/index.php)

[13] Flags of the World.com (http://www.fotw.us/flags/ch-ag014.html) accessed 28-January-2010

[14] Swiss inventory of cultural property of national and regional significance (http://www.bevoelkerungsschutz.admin.ch/internet/bs/de/home/themen/kgs/kgs_inventar.html) 21.11.2008 version, (German) accessed 28-Jan-2010

[15] ISOS site (http://www.isos.ch/en/nationale/index.asp) accessed 28-Jan-2010

External links

- Official webpage of the city of Baden (http://www.baden.ch)
- Schools in Baden (http://www.schule-baden.ch)
- *Baden* in German (http://www.hls-dhs-dss.ch/textes/d/D1633.php), French (http://www.hls-dhs-dss.ch/textes/f/F1633.php) and Italian (http://www.hls-dhs-dss.ch/textes/i/I1633.php) in the online *Historical Dictionary of Switzerland.*

La Tène culture

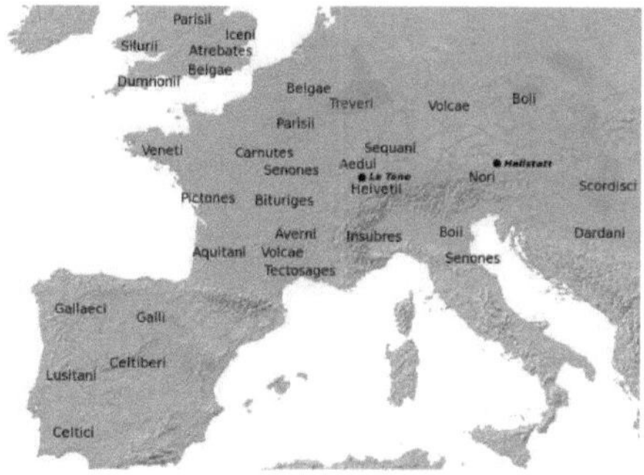

Overview of the Hallstatt and La Tène cultures. The core Hallstatt territory (800 BCE) is shown in solid yellow, the area of influence by 500 BCE (HaD) in light yellow. The core territory of the La Tène culture (450 BCE) is shown in solid green, the eventual area of La Tène influence by 50 BCE in light green. The territories of some major Celtic tribes are labelled.

Iron Age
↑ Bronze Age

Bronze Age collapse
Ancient Near East (1300 – 600 BC)
Aegean, Anatolia, Assyria, Caucasus, Cyprus, Egypt, Levant, Persia
India (1200 – 200)
Painted Grey Ware
Northern Black Polished Ware
Mauryan period
Anuradhapura Kingdom
Europe (1200 BC – AD 400)
Aegean
Caucasus
Novocherkassk
Hallstatt C
La Tène C
Villanovan C
British Iron Age
Greece, Rome, Celts
Scandinavia
China (600 BC – 200 BC)
Spring and Autumn Period and Warring States Period
Japan (100 BC – AD 500)
Yayoi period
Korea (400 BC – 50 BC)
Late Gojoseon period
Proto-Three Kingdoms period
Nigeria (400 BC – AD 200)
Axial Age Classical antiquity Zhou Dynasty Vedic period Alphabetic writing metallurgy
↓ Middle Ages
Historiography Greek, Roman, Chinese, Islamic

The **La Tène culture** was a European Iron Age culture named after the archaeological site of La Tène on the north side of Lake Neuchâtel in Switzerland, where a rich cache of artifacts was discovered by Hansli Kopp in 1857.

Reconstruction of a late La Tène period (1st century BCE) settlement in Altburg near Bundenbach.

La Tène culture developed and flourished during the late Iron Age (from 450 BCE to the Roman conquest in the 1st century BCE) in eastern France, Switzerland, Austria, southwest Germany, the Czech Republic, Poland, Slovakia, Slovenia, Hungary and Romania. To the north extended the contemporary Jastorf culture of Northern

Germany.[1] La Tène culture developed out of the early Iron Age Hallstatt culture without any definite cultural break, under the impetus of considerable Mediterranean influence from the Culture of Golasecca,[2] the Greeks in pre-Roman Gaul and later Etruscan civilizations.[3] Barry Cunliffe notes localisation of La Tène culture during the 5th century when there arose "two zones of power and innovation: a Marne – Moselle zone in the west with trading links to the Po Valley via the central Alpine passes and the Golasecca culture, and a Bohemian zone in the east with separate links to the Adriatic via the eastern Alpine routes and the Venetic culture".[4] A shift of settlement centres took place in the 4th century.

La Tène cultural material appeared over a large area, including parts of Ireland and Great Britain, northern Spain, Burgundy, and Austria. Elaborate burials also reveal a wide network of trade. In Vix, France, an elite woman of the 6th century BCE was buried with a bronze cauldron made in Greece. Exports from La Tène cultural areas to the Mediterranean cultures were based on salt, tin and copper, amber, wool and leather, furs and gold.

A disputed La Tène "homeland"

A 1st century BCE mirror found in Desborough, Northants, showing the spiral and trumpet theme.

Though there is no agreement on the precise region in which La Tène culture first developed, there is a broad consensus that the center of the culture lay on the northwest edges of Hallstatt culture, north of the Alps, within the region between the valleys of the Marne and Moselle in the west and modern Bavaria and Austria in the east. In 1994 a prototypical ensemble of elite grave sites of the early 5th century BCE was excavated at Glauberg in Hesse, northeast of Frankfurt-am-Main, in a region that had formerly been considered peripheral to the La Tène sphere.[5]

From their homeland, La Tène groups expanded in the 4th century to Hispania, the Po Valley, the Balkans, and even as far as Asia Minor, in the course of several major migrations. In the 4th century, a Gallic army led by Brennus reached Rome and took the city. In the 3rd century, Gallic bands entered Greece and threatened the oracle of Delphi, while another band settled Galatia in Asia Minor.

Periodization

Extensive contacts through trade are recognized in foreign objects deposited in elite burials; stylistic influences on La Tène material culture can be recognized in Etruscan, Italic, Greek, Dacian and Scythian sources. Dateable Greek pottery at La Tène sites and dendrochronology and thermoluminescence help provide date ranges in an absolute chronology at some La Tène sites.

As with many archaeological periods, La Tène history was originally divided into "early" (6th century BCE), "middle" (*ca* 450-100 BCE), and "late" (1st century BCE) stages, with the Roman occupation effectively driving the culture underground and ending its development. A broad cultural unity was not paralleled by overarching social-political unifying structures, and the extent to which the material culture can be linguistically linked is debated.

Ethnology

Our knowledge of this cultural area derives from three sources: from archaeological evidence, from Greek and Latin literary evidence, and more controversially, from ethnographical evidence suggesting some La Tène artistic and cultural survivals in traditionally Celtic regions of far western Europe. Some of the societies that are archaeologically identified with La Tène material culture were identified by Greek and Roman authors from the 5th century onwards as *Keltoi* ("Celts") and *Galli* ("Gauls"). Herodotus (iv.49) correctly placed *Keltoi* at the source of the Ister/Danube, in the heartland of La Tène material culture: "The Ister flows right across Europe, rising in the country of the Celts", whom however, apparently misunderstanding his source,[6] he places "farthest to the west of any people of Europe"[7] Whether the usage of classical sources means that the whole of La Tène culture can be attributed to a unified Celtic people is difficult to assess; archaeologists have repeatedly concluded that language, material culture, and political affiliation do not necessarily run parallel. Frey notes (Frey 2004) that in the 5th century, "burial customs in the Celtic world were not uniform; rather, localised groups had their own beliefs, which, in consequence, also gave rise to distinct artistic expressions". In some cases where La Tène archaeological sites are overlain by Slavic culture, any identification of La Tène material culture with Celts may become a sensitive local issue.

Material culture

La Tène metalwork in bronze, iron and gold, developing technologically out of Halstatt culture, is stylistically characterized by inscribed and inlaid intricate spirals and interlace, on fine bronze vessels, helmets and shields, horse trappings and elite jewelry, especially the neck rings called torcs and elaborate clasps called *fibulae*. It is characterized by elegant, stylized curvilinear animal and vegetal forms, allied with the Hallstatt traditions of geometric patterning. The Early Style of La Tène art and culture mainly featured static, geometric decoration, while the transition to the Developed Style constituted a shift to movement-based forms, such as triskeles. Some subsets within the Developed Style contain more specific design trends, such as the recurrent serpentine scroll of the Waldalgesheim Style [8]

Initially La Tène folk lived in open settlements that were dominated by the chieftains' towering hill forts.[9] The development of towns— *oppida*— appears in mid-La Tène culture. La Tène dwellings were carpenter-built rather than of masonry. La Tène peoples also dug ritual shafts, in which votive offerings and even human sacrifices were cast. Severed heads appear to have held great power and were often represented in carvings. Burial sites included weapons, carts, and both elite and household goods, evoking a strong continuity with an afterlife.

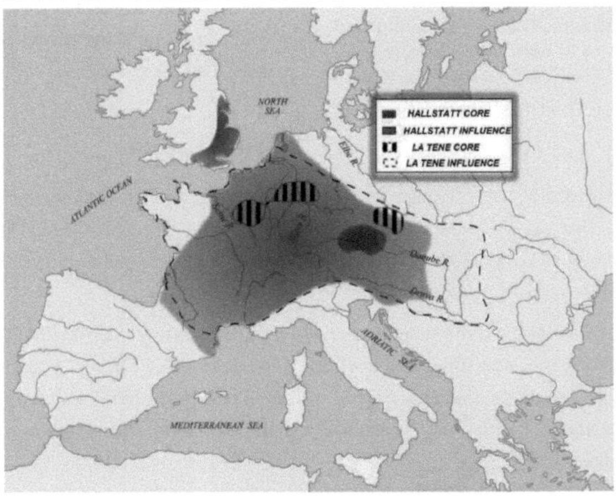

The *Hallstatt* and *La Tène* Cultures

Discovery

La Tène is a village on the northern shore of Lake Neuchâtel, Switzerland. It is both an archaeological site and the eponymous site for the late Iron Age La Tène culture, also spelt "Latène" or "La-Tène".

In 1857, prolonged drought lowered the waters of the lake by about 2 m. On the northernmost tip of the lake, between the river Thièle and a point south of the village of Marin-Epagnier, Hansli Kopp, looking for antiquities for Colonel Frédéric Schwab, discovered several rows of wooden piles that still reached about 50 cm into the water. From among these, Kopp collected about forty iron swords.

The Swiss archaeologist Ferdinand Keller published his findings in 1868 in his influential first report on the Swiss pile dwellings (*Pfahlbaubericht*). In 1863 he interpreted the remains as a Celtic village built on piles. Eduard Desor, a geologist from Neuchâtel, started excavations on the lakeshore soon afterwards. He interpreted the site as an armory, erected on piles over the lake and later destroyed by enemy action. Another interpretation accounting for the presence of cast iron swords that had not been sharpened, was of a site of sacrifices.

With the first systematic lowering of the Swiss lakes from 1868 to 1883, the site fell completely dry. In 1880, Emile Vouga, a teacher from Marin-Epagnier, uncovered the wooden remains of two bridges (designated "Pont Desor" and "Pont Vouga") originally over 100 m long, that crossed the little Thièle River (today a nature reserve) and the remains of five houses on the shore. After Vouga had finished, F. Borel, curator of the Marin museum, began to excavate as well. In 1885 the canton asked the Société d'Histoire of Neuchâtel to continue the excavations, the results of which were published by Vouga in the same year.

All in all, over 2500 objects, mainly made from metal, have been excavated in La Tène. Weapons predominate, there being 166 swords (most without traces of wear), 270 lanceheads, and 22 shield bosses, along with 385 brooches, tools, and parts of chariots. Numerous human and animal bones were found as well.

Interpretations of the site vary. Some scholars believe the bridge was destroyed by high water, while others see it as a place of sacrifice after a successful battle (there are almost no female ornaments).

An exhibition marking the 150th anniversary of the discovery of the La Tène site was launched in June 2007 at the Musée Schwab in Bienne, Switzerland. It is scheduled to move to Zürich in 2008 and the Mont Beuvray in Burgundy in 2009.

Sites

Some sites are:

- La Tène, Marin-Epagnier
- Bern, Engehalbinsel: oppidum
- Jolimont
- Manching: oppidum
- Mormont
- Münsingen, burial field
- Petinesca
- Basel oppidum
- Bibracte, oppidum of the Aedui at Mont Beuvray in Burgundy
- Erstfeld hoard
- Hochdorf Chieftain's Grave
- Bopfingen: *Viereckschanze*, a characteristic rectangular enclosure
- Fellbach-Schmiden, near Stuttgart: *Viereckschanze*; ritual objects recovered from a well
- Kleinaspergle: elite graves of La Tène I
- Waldalgesheim: an elite chariot burial, 4th century
- Glauberg, oppidum and elite graves
- Dürrnberg near Hallein: Burial field and earthworks of late Hallstatt–early La Tène
- Donnersberg: oppidum
- Vill near Innsbruck: remains of dwellings
- Sandberg Celtic city near Platt and Roseldorf in Lower Austria

- Vix/Mont Lassois: oppidum and elaborate graves
- Titelberg: oppidum in Luxembourg
- Reinheim: Tomb of a princess/priestess with burial gifts

Artifacts

Some outstanding La Tène artifacts are:

Vix krater

- "Strettweg Cart" (7th century BCE), found in southeast Austria, a four-wheeled cart with a goddess, riders with axes and shields, attendants and stags. (Landesmuseum Johanneum, Graz, Austria)
- A woman in Vix (Châtillon-sur-Seine, Burgundy) buried with a 1100 litre (290 gallon) bronze Greek vase, the largest ever found.
- The silver "Gundestrup cauldron" (3rd or 2nd century BCE), found ritually broken in a peat bog near Gundestrup, Denmark, but probably made near the Black Sea, perhaps in Thrace. (National Museum of Denmark, Copenhagen)
- "Battersea Shield" (350-50 BCE), found in the Thames, made of bronze with red enamel. (British Museum, London)
- "Witham Shield" (4th century BCE). (British Museum, London) [10] [11]
- "Chertsey Shield (400-200 BCE). (British Museum, London) [12]
- "Turoe stone" in Galway, Ireland
- Chariot burial found at Waldalgesheim, Bad Kreuznach, Germany, late 4th century BCE. (Bonn: Rheinisches Landesmuseum)
- Chariot burial found at La Gorge Meillet (St-Germain-en-Laye: Musée des Antiquités Nationales).
- A life-sized sculpture of a warrior that accompanied the Glauberg burials.
- A gold-and-bronze model of an oak tree (3rd century BCE) found at the Oppidum of Manching.
- Noric steel

Notes

[1] The Jastorf culture is the southern part of the Pre-Roman Iron Age of the north.

[2] Venceslas Kruta, *La grande storia dei Celti. La nascita, l'affermazione, la decadenza*, (Newton & Compton), Roma, 2003 ISBN 9788882898519 , a translation of *Les Celtes, histoire et dictionnaire. Des origines à la romanisation et au christianisme*, Robert Laffont, Paris, 2000, without the dictionary;

[3] Sarunas Milisauskas , *European prehistory: a survey* p.354 (http://books.google.com/books?id=31LFIITb3LUC&pg=PA354)

[4] Cunliffe 1997:66.

[5] Mystery of the Celts (http://www.frankfurtlounge.de/Mystery_of_the_Celts.htm).

[6] Lionel Pearson, "Herodotus on the Source of the Danube", *Classical Philology* **29**.4 (October 1934:328-337).

[7] In another place (ii.33.) Herodotus mentions the Ister, which "rising in the country of the Celts, beginning from the city of Pyrene, cuts Europe in half", which would have made it intersect with the Rhone; *Pyrene* is not mentioned aside from this context.

[8] Harding, D.W. The Archaeology of Celtic Art. New York: Routledge, 2007. Print.

[9] frankfurt,kultur,highlights,mystery of the kelts (http://www.frankfurtlounge.de/Mystery_of_the_Celts.htm)

[10] British Museum - The Witham Shield (http://www.britishmuseum.org/explore/highlights/highlight_objects/pe_prb/t/the_witham_shield.aspx)

[11] Treasures Of The Celtic Causeway - The Fiskerton Log Boat | Culture24 (http://www.24hourmuseum.org.uk/nwh/art12215.html)

[12] British Museum - Chertsey Shield (http://www.britishmuseum.org/explore/highlights/highlight_objects/pe_prb/c/chertsey_shield.aspx)

Further reading

- Cunliffe, Barry. *The Ancient Celts*. Oxford: Oxford University Press. 1997
- Collis, John. *The Celts: Origins, Myths, Invention*. London: Tempus, 2003.
- Kruta, Venceslas, *La grande storia dei Celti. La nascita, l'affermazione, la decadenza*, Newton & Compton, Roma, 2003 ISBN 9788882898519 (492 pagine - traduzione di *Les Celtes, histoire et dictionnaire. Des origines à la romanisation et au Christianisme*, Robert Laffont, Paris, 2000, con esclusione del dizionario)
- James, Simon. *The Atlantic Celts*. London: British Museum Press, 1999.
- James, Simon, and Valery Rigby. *Britain and the Celtic Iron Age*. London: British Museum Press, 1997.

See also

- Golasecca culture

External links

- Excellent and detailed summary of excavations and interpretations (http://www.archaeologisch.de/forschung/latene/latene.html) (in German)
- La Tène site (http://www.edunet.ch/classes/marin/la_tene.htm): brief text, illustrations (in French)
- Images from World History: Iron Age Western Europe (http://www.hp.uab.edu/image_archive/uj/ujk.html)
- Charles Bergengren, Cleveland Institute of Art, 1999: (http://gate.cia.edu/cbergengren/arthistory/celtic/slides.html) illustrations of La Tène artifacts
- Otto Hermann Frey, "A new approach to early Celtic art" (http://www.ria.ie/publications/journals/journaldb/index.asp?select=fulltext&id=100427.). Setting the Glauberg finds in context of shifting iconography.
- Images from the Logan Museum of Anthropology at Beloit College (http://www.beloit.edu/logan/exhibitions/virtual_exhibitions/before_history/europe/station_la_tene.php)
- La Tène Archaeological Sites in Romania (http://www.eliznik.org.uk/RomaniaHistory/romania-arch-map/latene.htm)

Aargau

Kanton Aargau	
— Canton of Switzerland —	
 Coat of arms	

Coordinates: 47°5′N 8°0′E	
Capital	Aarau
Largest City	Wettingen
Subdivisions	220 municipalities, 11 districts
Government	
- Executive	Executive Council (5)
- Legislative	Grand Council (140)
Area	
- Total	1404 km^2 (542.1 sq mi)
Population (12/2010)[1]	
- Total	612611
- Density	436.3/km^2 (1130.1/sq mi)
Highest point	908 m (2979 ft) - Geissfluegrat
Lowest point	260 m (853 ft) - Rhine at Kaiseraugst
Joined	1803
Abbreviation	AG
Languages	German
Website	AG.ch [2]

Aargau (German **Aargau**; rarely anglicized **Argovia**) is one of the more northerly cantons of Switzerland. It comprises the lower course of the river Aare, which is why the canton is called Aar-gau (meaning *Aare district*).

History

Aarau

In early medieval times, *Argovia* or *Argowe* was a disputed border region between the duchies of Alamannia and Burgundy. A line of the von Wetterau (Conradines) intermittently held the countship of Aargau from 750 until about 1030, when they lost it (having in the meantime taken the name von Tegerfelden). From the extinction in 1254 of the Hohenstaufen dynasty until 1415, the area was ruled by the Habsburgs, and many castles from that time still stand (examples include Habsburg, Lenzburg, Tegerfelden, Bobikon, Stin and Wildegg). The Habsburgs founded a number of monasteries (with some structures enduring, e.g., in Wettingen and Muri), the closing of which by the government in 1841 was a contributing factor to the outbreak of the Swiss civil war – the "Sonderbund War" – in 1847.

In 1415, the Aargau was taken from the Habsburgs by the Swiss Confederates. Bern kept the southwest portion (Zofingen, Aarburg, Aarau, Lenzburg, and Brugg). Some districts, named the *Freie Ämter* (*free bailiwicks*) – Mellingen, Muri, Villmergen, and Bremgarten), with the countship of Baden – were governed as "subject lands" by all or some of the Confederates.

French forces occupied Aargau from 10 March to 18 April 1798; thereafter the Bernese portion became the canton of Aargau in the Helvetic Republic and the remainder formed the Canton of Baden. In 1803, the two halves were united under the name of canton of Aargau,[3] which was then admitted as a full member of the reconstituted Confederation. Fricktal, ceded in 1802 by Austria via Napoleonic France to the Helvetic Republic, was briefly a separate Swiss canton under a *Statthalter* ('Lieutenant'), but on 9 March 1803 was incorporated in the canton of Aargau.

The chief magistracy of Aargau changed its style repeatedly:

- first two consecutive *Regierungsstatthalter* :
  - April 1798 – November 1801 Jakob Emmanuel Feer (b. 1754 – d. 1833)
  - 1802 – 1803 Johann Heinrich Rothpletz (b. 1766 – d. 1833)
- Presidents of the Government Commission
  - 10 March 1803 – 26 April 1803 Johann Rudolf Dolder (b. 1753 – d. 1807)
  - 26 April 1803 – 1815 a 'Small Council' (president rotating monthly)
- annual *Amtsbürgermeister* 1815–1831
- annual *Landammänner* since 1815

In the year 2003, the canton of Aargau celebrated its 200th anniversary.

In the 17th Century, Jews were banished from Switzerland. However, a few families were permitted to live in two villages, Endingen and Lengnau, in Aargau which became the Jewish ghetto in Switzerland. This remained the case until the 19th Century. In 1799 all special tolls were abolished, and in 1802 the poll tax was removed. On 5 May, 1809, the right of citizenship was granted to Jews, and they were permitted to engage in trade and agriculture. The right of settlement remained restricted to Endingen and Lengnau until 7 May 1846, when they were allowed to settle in any portion of the canton of Aargau. Ten years later (24 September 1856) the Swiss Federal Council voted them equal political rights with other Swiss citizens in that canton, as well as entire freedom of commerce; but the opposition of the Christian population prevented the decision from being generally carried out. The federal authorities in July 1863, granted the Swiss Jews the fullest rights of citizens. However, full civil equality was obtained only when they received the formal rights of citizenship, which had long been withheld from them in their own communities of Endingen and Lengnau. A resolution of the Grand Council, on 15 May 1877, granted citizens' rights to the members of the Jewish communities of those places, giving them charters under the names of New Endingen and New Lengnau.[4]

Geography

View of the Lägern from Bözberg

Its total area is 1404 square kilometers (542 sq mi). The capital is Aarau. It borders Germany to the north. To the west lie the cantons of Basel-Landschaft, Solothurn and Bern. The canton of Lucerne lies south of the canton of Aargau, Zürich and Zug to the east.

The canton of Aargau is one of the least mountainous Swiss cantons, forming part of a great table-land, to the north of the Alps and the east of the Jura, above which rise low hills. The surface of the country is beautifully diversified, undulating tracts and well-wooded hills[5] alternating with fertile valleys watered mainly by the Aar and its tributaries. The valleys alternate with pleasant hills, most of which are full of woods. Slightly over one-third of the canton is wooded (518 square kilometers (200 sq mi)), while nearly half is used from farming (635.7 square kilometers (245.4 sq mi)). 33.5 square kilometers (12.9 sq mi) or about 2.4% of the canton is considered unproductive, mostly lakes and streams. The canton is heavily developed, with 216.7 square kilometers (83.7 sq mi) or about 15% of the canton developed for housing or transportation.[6]

It contains the famous hot sulphur springs of Baden and Schinznach-Bad, while at Rheinfelden there are very extensive saline springs. Just below Brugg the Reuss and the Limmat join the Aar, while around Brugg are the ruined castle of Habsburg, the old convent of Königsfelden (with fine painted medieval glass) and the remains of the Roman settlement of *Vindonissa* (Windisch).

Political subdivisions

Districts

Districts in Aargau

Aargau is divided into 11 districts:

- Aarau with capital Aarau
- Baden with capital Baden
- Bremgarten with capital Bremgarten
- Brugg with capital Brugg
- Kulm with capital Unterkulm
- Laufenburg with capital Laufenburg
- Lenzburg with capital Lenzburg
- Muri with capital Muri
- Rheinfelden with capital Rheinfelden
- Zofingen with capital Zofingen
- Zurzach with capital Zurzach

Municipalities

There are a total of 231 municipalities in the canton of Aargau.

Demographics

The population of Aargau is 612611 (31 December 2010).[1] As of 2007 the population included 118,407 foreigners, who made up 20.36% of the total population.[7] Religiously, they are (as of 2000) nearly evenly split between Roman Catholics (40%) and Protestants (37%).[8] German is the most common language in the canton, spoken by 87% of the population.[9]

Economy

The farmland of the canton of Aargau is some of the most fertile in Switzerland. Dairy farming, cereal and fruit farming are among the canton's main economic activities. The canton is also industrially developed, particularly in the fields of electrical engineering, precision instruments, iron, steel and cement.

Leibstadt Nuclear Power Plant

Three of Switzerland's five nuclear power plants are in the canton of Aargau (Beznau I + II and Leibstadt). Additionally, the many rivers supply enough water for numerous hydroelectric power plants throughout the canton. The canton of Aargau is often called "the energy canton".

A significant number of people commute into the financial center of the city of Zürich, which is just across the cantonal border. As such the per capita cantonal income is approximately $41,777.[10]

Tourism is significant, particularly for the hot springs at Baden and Schinznach-Bad, the ancient castles, the landscape, and the many old museums in the canton.[5] Hillwalking is another tourist attraction but is of only limited significance.

See also

- Canton of Aargau castles and fortresses
- Aargauer Zeitung
- FC Aarau
- GP Kanton Aargau, bicycle race

References

[1] Canton of Aargau Statistical Office (http://www.ag.ch/staag/daten/B01/index.html), MS Excel document – *Bevölkerungsbestand per 31.12.2010 nach Gemeinde, Geschlecht u. Nationalität* (German) accessed 11 May 2011

[2] http://www.AG.ch

[3] Bridgwater, W. & Beatrice Aldrich. (1966) *The Columbia-Viking Desk Encyclopedia*. Columbia University. p. 11.

[4] Jewish Encyclopedia: Aargau (http://www.jewishencyclopedia.com/view.jsp?artid=3&letter=A)

[5] "Aargau" (http://search.eb.com/eb/article-9003207) *Encyclopædia Britannica*. 11 January 2007

[6] Federal Department of Statistics (2008). "Arealstatistik – Kantonsdaten nach 15 Nutzungsarten" (http://www.bfs.admin.ch/bfs/portal/de/index/themen/02/03/blank/key/01/zustand_und_entwicklung__tabelle.html) (Microsoft Excel).. Retrieved 2009-01-15.

[7] Federal Department of Statistics (2008). "Ständige Wohnbevölkerung nach Staatsangehörigkeit, Geschlecht und Kantonen" (http://www.bfs.admin.ch/bfs/portal/de/index/themen/01/02/blank/key/raeumliche_verteilung/kantone__gemeinden.html) (Microsoft Excel).. Retrieved 2008-11-05.

[8] Federal Department of Statistics (2004). "Wohnbevölkerung nach Religion" (http://www.bfs.admin.ch/bfs/portal/de/index/themen/01/05/blank/key/religionen.html) (Interactive Map).. Retrieved 2009-01-15.

[9] Federal Department of Statistics (2004). "Wohnbevölkerung nach Sprachen" (http://www.bfs.admin.ch/bfs/portal/de/index/themen/01/05/blank/key/sprachen.html) (Interactive Map). . Retrieved 2009-01-16.
[10] http://www.bfs.admin.ch/bfs/portal/en/index/themen/04/02/05/key/pro-kopf-einkommen.html

External links

- Official website (http://www.ag.ch) **(German)**
- *Aargau* in German (http://www.hls-dhs-dss.ch/textes/d/D7392.php), French (http://www.hls-dhs-dss.ch/textes/f/F7392.php) and Italian (http://www.hls-dhs-dss.ch/textes/i/I7392.php) in the online *Historical Dictionary of Switzerland.*
- Swiss Federal Statistical Office-Aargau Statistics (http://www.bfs.admin.ch/bfs/portal/en/index/regionen/regionalportraets/aargau/blank/kennzahlen.html)

Switzerland

Motto: *(unofficial)* "Unus pro omnibus, omnes pro uno" (Latin)
English: *One for all, all for one*
German: *Einer für alle, alle für einen*
French: *Un pour tous, tous pour un*
Italian: *Uno per tutti, tutti per uno*
Romansh: *In per tuts, tuts per in*

Anthem: Swiss Psalm
Schweizerpsalm (German)
Cantique Suisse (French)
Salmo svizzero (Italian)
Psalm svizzer (Romansch)

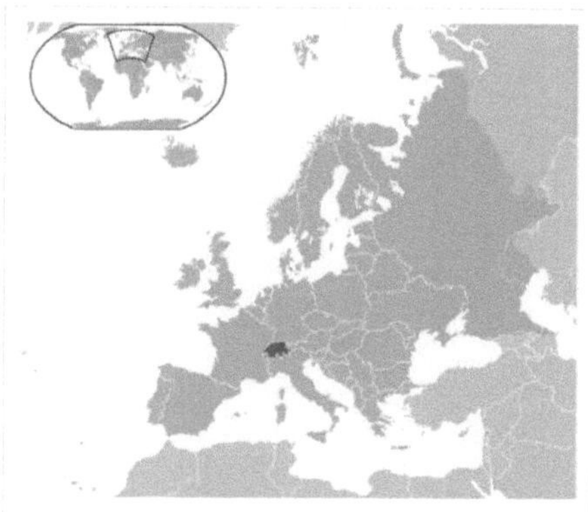

Location of Switzerland(green)
in Europe(dark grey) — [Legend]

Capital	Bern[1] (de facto) 46°57′N 7°27′E
Largest city	Zurich
Official language(s)	German (63.7 %), French (20.4 %), Italian (6.5 %), Romansh[2] (0.5 %)
Demonym	Swiss
Government	Federal republic, with directorial system and direct democracy

-	Federal Council	Micheline Calmy-Rey (Pres. 11) Doris Leuthard Eveline Widmer-Schlumpf (VP 11) Ueli Maurer Didier Burkhalter Simonetta Sommaruga Johann Schneider-Ammann
-	Federal Chancellor	Corina Casanova
Legislature		Federal Assembly
-	Upper House	Council of States
-	Lower House	National Council
Independence		
-	Foundation date	1 August[3] 1291
-	*de facto*	22 September 1499
-	Recognised	24 October 1648
-	Restored	7 August 1815
-	Federal state	12 September 1848[4]
Area		
-	Total	41,285 km^2 (133rd) 15,940 sq mi
-	Water (%)	4.2
Population		
-	2010 estimate	7,866,500[5] (95th)
-	2000 census	7,452,075
-	Density	188/km^2 (65th) 477.4/sq mi
GDP (PPP)		2011 estimate
-	Total	$321.898 billion[6]
-	Per capita	$45,265[6]
GDP (nominal)		2011 estimate
-	Total	$512.065 billion[6]
-	Per capita	$75,835[6]
Gini (2000)		33.7 (medium)
HDI (2010)		▲ 0.874[7] (very high) (13th)
Currency		Swiss franc (CHF)
Time zone		CET (UTC+1)
-	Summer (DST)	CEST (UTC+2)
Drives on the		right (trains: left)
ISO 3166 code		CH
Internet TLD		.ch

Calling code	+41

Switzerland (German: *die Schweiz,*[8] French: *Suisse*, Italian: *Svizzera*, Romansh: *Svizra*), officially the **Swiss Confederation** (Latin: *Confoederatio Helvetica*, hence its abbreviation CH), is a federal republic consisting of 26 cantons, with Bern as the seat of the federal authorities. The country is situated in Western Europe,[9] where it is bordered by Germany to the north, France to the west, Italy to the south, and Austria and Liechtenstein to the east.

Switzerland is a landlocked country geographically divided between the Alps, the Central Plateau and the Jura, spanning an area of 41285 km^2 (15940 sq mi). While the Alps occupy the greater part of the territory, the Swiss population of approximately 7.9 million people concentrates mostly on the Plateau, where the largest cities are to be found. Among them are the two global cities and economic centres of Zurich and Geneva.

The Swiss Confederation has a long history of neutrality—it has not been in a state of war internationally since 1815—and did not join the United Nations until 2002. It pursues, however, an active foreign policy and is frequently involved in peace-building processes around the world.[10] Switzerland is also the birthplace of the Red Cross and home to a large number of international organizations, including the second largest UN office. On the European level, it is a founding member of the European Free Trade Association and is part of the Schengen Area – although it is notably not a member of the European Union, nor the European Economic Area.

In nominal terms, Switzerland is one of the richest countries in the world by per capita gross domestic product, with a nominal per capita GDP of $75,835.[6] In 2010, Switzerland had the highest wealth per adult of any country in the world (with $372,692 for each person).[11] Switzerland also has one of the world's largest account balances as a percentage of GDP, only placing behind a few oil producing countries. Zurich and Geneva have respectively been ranked as the cities with the second and third highest quality of life in the world.[12] In 2010 the World Economic Forum ranked Switzerland as the most competitive country in the world,[13] while ranked by the European Union as Europe's most innovative country by far.[14]

Switzerland comprises three main linguistic and cultural regions: German, French, and Italian, to which the Romansh-speaking valleys are added. The Swiss therefore do not form a nation in the sense of a common ethnic or linguistic identity. The strong sense of belonging to the country is founded on the common historical background, shared values (federalism, direct democracy, neutrality)[15] and Alpine symbolism.[16] The establishment of the Swiss Confederation is traditionally dated to 1 August 1291; Swiss National Day is celebrated on the anniversary.

Etymology

The English name *Switzerland* is a compound containing *Switzer*, an obsolete term for the Swiss, which was in use during the 16th to 19th centuries.[17] The English adjective *Swiss* is a loan from French *Suisse*, also in use since the 16th century. The name *Switzer* is from the Alemannic *Schwiizer*, in origin an inhabitant of *Schwyz* and its associated territory, one of the Waldstätten cantons which formed the nucleus of the Old Swiss Confederacy. The toponym itself is first attested in 972, as Old High German *Suittes*, ultimately perhaps related to *suedan* "to burn", referring to the area of forest that was burned and cleared to build.[18] The name was extended to the area dominated by the canton, and after the Swabian War of 1499 gradually came to be used for the entire Confederation.[19] [20]

The Swiss German name of the country, *Schwiiz*, is homophonous to that of the canton and the settlement, but distinguished by the use of the definite article (*d'Schwiiz* for the Confederation,[21] but simply *Schwiiz* for the canton and the town).[22]

The New Latin name *Confoederatio Helvetica* was introduced gradually after the formation of the federal state in 1848, harking back to the Napoleonic Helvetic Republic, appearing on coins from 1879, inscribed on the Federal Palace in 1902 and after 1948 used in the official seal.[23] It is derived from the name of the *Helvetii*, a Gaulish tribe living on the Swiss plateau before the Roman era. *Helvetia* appears as a national personification of the Swiss confederacy in the 17th century, with a 1672 play by Johann Caspar Weissenbach.[24]

History

Switzerland has existed as a state in its present form since the adoption of the Swiss Federal Constitution in 1848. The precursors of modern Switzerland established a protective alliance at the end of the 13th century (1291), forming a loose confederation of states which persisted for centuries.

Early history

The oldest traces of hominid existence in Switzerland date back about 150,000 years.[25] The oldest known farming settlements in Switzerland, which were found at Gächlingen, have been dated to around 5300 BC.[25]

Founded in 44 BC, Augusta Raurica was the first Roman settlement on the Rhine and is now the most important archaeological site in Switzerland[26]

The earliest known cultural tribes of the area were members of the Hallstatt and La Tène cultures, named after the archaeological site of La Tène on the north side of Lake Neuchâtel. La Tène culture developed and flourished during the late Iron Age from around 450 BC,[25] possibly under some influence from the Greek and Etruscan civilisations. One of the most important tribal groups in the Swiss region was the Helvetii. In 58 BC, at the Battle of Bibracte, Julius Caesar's armies defeated the Helvetii.[25] In 15 BC, Tiberius, who was destined to be the second Roman emperor and his brother, Drusus, conquered the Alps, integrating them into the Roman Empire. The area occupied by the Helvetii—the namesakes of the later *Confoederatio Helvetica*—first became part of Rome's Gallia Belgica province and then of its Germania Superior province, while the eastern portion of modern Switzerland was integrated into the Roman province of Raetia.

In the Early Middle Ages, from the 4th century, the western extent of modern-day Switzerland was part of the territory of the Kings of the Burgundians. The Alemanni settled the Swiss plateau in the 5th century and the valleys of the Alps in the 8th century, forming Alemannia. Modern-day Switzerland was therefore then divided between the kingdoms of Alemannia and Burgundy.[25] The entire region became part of the expanding Frankish Empire in the 6th century, following Clovis I's victory over the Alemanni at Tolbiac in 504 AD, and later Frankish domination of the Burgundians.[27] [28]

Throughout the rest of the 6th, 7th and 8th centuries the Swiss regions continued under Frankish hegemony (Merovingian and Carolingian dynasties). But after its extension under Charlemagne, the Frankish empire was divided by the Treaty of Verdun in 843.[25] The territories of present day Switzerland became divided into Middle Francia and East Francia until they were reunified under the Holy Roman Empire around 1000 AD.[25]

By 1200, the Swiss plateau comprised the dominions of the houses of Savoy, Zähringer, Habsburg and Kyburg.[25] Some regions (Uri, Schwyz, Unterwalden, later known as *Waldstätten*) were accorded the Imperial immediacy to grant the empire direct control over the mountain passes. When the Kyburg dynasty fell in 1264 AD, the Habsburgs under King Rudolph I (Holy Roman Emperor in 1273) extended their territory to the eastern Swiss plateau.[27]

Old Swiss Confederacy

Further information: Growth of the Old Swiss Confederacy, Reformation in Switzerland, and Early Modern Switzerland

The Old Swiss Confederacy was an alliance among the valley communities of the central Alps. The Confederacy facilitated management of common interests (free trade) and ensured peace on the important mountain trade routes The Federal Charter of 1291 agreed between the rural communes of Uri, Schwyz, and Unterwalden is considered the confederacy's founding document, even though similar alliances are likely to have existed decades earlier.[29] [30]

Federal charter of 1291

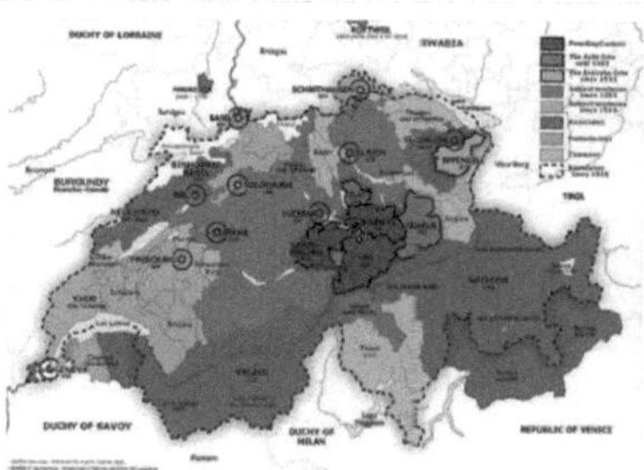

The Old Swiss Confederacy from 1291 (dark grey) to the sixteenth century (light grey) and its associates (blue)

By 1353, the three original cantons had joined with the cantons of Glarus and Zug and the Lucerne, Zurich and Bern city states to form the "Old Confederacy" of eight states that existed until the end of the 15th century. The expansion led to increased power and wealth for the federation.[30] By 1460, the confederates controlled most of the territory south and west of the Rhine to the Alps and the Jura mountains, particularly after victories against the Habsburgs (Battle of Sempach, Battle of Näfels), over Charles the Bold of Burgundy during the 1470s, and the success of the Swiss mercenaries. The Swiss victory in the Swabian War against the Swabian League of Emperor Maximilian I in 1499 amounted to *de facto* independence within the Holy Roman Empire.[30]

The Old Swiss Confederacy had acquired a reputation of invincibility during these earlier wars, but expansion of the federation suffered a setback in 1515 with the Swiss defeat in the Battle of Marignano. This ended the so-called "heroic" epoch of Swiss history.[30] The success of Zwingli's Reformation in some cantons led to inter-cantonal religious conflicts in 1529 and 1531 (Wars of Kappel). It was not until more than one hundred years after these internal wars that, in 1648, under the Peace of Westphalia, European countries recognized Switzerland's independence from the Holy Roman Empire and its neutrality.[27] [28]

During the Early Modern period of Swiss history, the growing authoritarianism of the patriciate families combined with a financial crisis in the wake of the Thirty Years' War led to the Swiss peasant war of 1653. In the background to this struggle, the conflict between Catholic and Protestant cantons persisted, erupting in further violence at the Battles of Villmergen in 1656 and 1712.[30]

Napoleonic era

In 1798, the revolutionary French government conquered Switzerland and imposed a new unified constitution.[30] This centralised the government of the country and effectively abolished the cantons and Mülhausen and Valtellina valley separated from Switzerland. The new regime, known as the Helvetic Republic, was highly unpopular. It had been imposed by a foreign invading army and destroyed centuries of tradition, making Switzerland nothing more than a French satellite state. The fierce French suppression of the Nidwalden Revolt in September 1798 was an example of the oppressive presence of the French Army and the local population's resistance to the occupation.

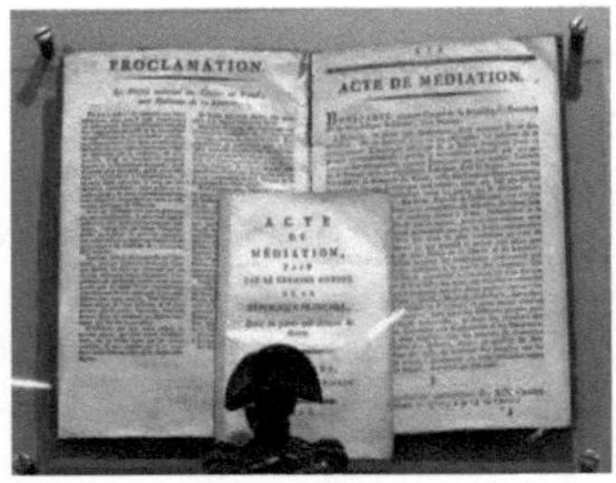

The Act of Mediation was Napoleon's attempt at a compromise between the Ancien Régime and a Republic.

When war broke out between France and its rivals, Russian and Austrian forces invaded Switzerland. The Swiss refused to fight alongside the French in the name of the Helvetic Republic. In 1803 Napoleon organised a meeting of the leading Swiss politicians from both sides in Paris. The result was the Act of Mediation which largely restored Swiss autonomy and introduced a Confederation of 19 cantons.[30] Henceforth much of Swiss politics would concern balancing the cantons' tradition of self-rule with the need for a central government.

In 1815 the Congress of Vienna fully re-established Swiss independence and the European powers agreed to permanently recognise Swiss neutrality.[27] [28] [30] Swiss troops still served foreign governments until 1860 when they fought in the Siege of Gaeta. The treaty also allowed Switzerland to increase its territory, with the admission of the cantons of Valais, Neuchâtel and Geneva. Switzerland's borders have not changed since.[31]

Federal state

The restoration of the power to the patriciate was only temporary. After a period of unrest with repeated violent clashes such as the Züriputsch of 1839, civil war broke out in 1847 when some of the Catholic cantons tried to set up a separate alliance (the *Sonderbundskrieg*).[30] The war lasted for less than a month, causing fewer than 100 casualties, most of which were through friendly fire. However minor the Sonderbundskrieg seems to be when compared with other European riots and wars in the 19th century, it nevertheless had a major impact on both the psychology and the society of the Swiss and of Switzerland.

The first Federal Palace in Bern (1857). One of the three cantons presiding over the Tagsatzung (former legislative and executive council), Bern was chosen as the federal capital in 1848, mainly because of its closeness to the French speaking area.[32]

The war convinced most Swiss of the need for unity and strength towards its European neighbours. Swiss people from all strata of society, whether Catholic, Protestant, or from the liberal or conservative current, realised that the cantons would profit more if their economic and religious interests were merged.

Thus, while the rest of Europe was plagued by revolutionary uprisings, the Swiss drew up a constitution which provided for a federal layout, much of it inspired by the American example. This constitution provided for a central authority while leaving the cantons the right to self-government on local issues. Giving credit to those who favoured the power of the cantons (the Sonderbund Kantone), the national assembly was divided between an upper house (the Swiss Council of States, 2 representatives per canton) and a lower house (the National Council of Switzerland, representatives elected from across the country). Referenda were made mandatory for any amendment of this constitution.[28]

Inauguration in 1882 of the Gotthard Rail Tunnel connecting the southern canton of Ticino, the longest in the world at the time.[33]

A system of single weights and measures was introduced and in 1850 the Swiss franc became the Swiss single currency. Article 11 of the constitution forbade sending troops to serve abroad, though the Swiss were still obliged to serve Francis II of the Two Sicilies with Swiss Guards present at the Siege of Gaeta in 1860, marking the end of foreign service.

An important clause of the constitution was that it could be re-written completely if this was deemed necessary, thus enabling it to evolve as a whole rather than being modified one amendment at a time.[34]

This need soon proved itself when the rise in population and the Industrial Revolution that followed led to calls to modify the constitution accordingly. An early draft was rejected by the population in 1872 but modifications led to its acceptance in 1874.[30] It introduced the facultative referendum for laws at the federal level. It also established federal responsibility for defense, trade, and legal matters.

In 1891, the constitution was revised with unusually strong elements of direct democracy, which remain unique even today.[30]

Modern history

General Ulrich Wille, Commander-in-Chief of the Swiss Army during World War I

Switzerland was not invaded during either of the world wars. During World War I, Switzerland was home to Vladimir Illych Ulyanov (Lenin) and he remained there until 1917.[35] Swiss neutrality was seriously questioned by the Grimm-Hoffmann Affair in 1917, but it was short-lived. In 1920, Switzerland joined the League of Nations, which was based in Geneva, on the condition that it was exempt from any military requirements.

During World War II, detailed invasion plans were drawn up by the Germans,[36] but Switzerland was never attacked.[30] Switzerland was able to remain independent through a combination of military deterrence, concessions to Germany, and good fortune as larger events during the war delayed an invasion.[28] [37] Under General Henri Guisan, a massive mobilisation of militia forces was ordered. The Swiss military strategy was changed from one of static defence at the borders to protect the economic heartland, to one of organised long-term attrition and withdrawal to strong, well-stockpiled positions high in the Alps known as the Reduit. Switzerland was an important base for espionage by both sides in the conflict and often mediated communications between the Axis and Allied powers.[37]

Switzerland's trade was blockaded by both the Allies and by the Axis. Economic cooperation and extension of credit to the Third Reich varied according to the perceived likelihood of invasion and the availability of other trading partners. Concessions reached a peak after a crucial rail link through Vichy France was severed in 1942, leaving Switzerland completely surrounded by the Axis. Over the course of the war, Switzerland interned over 300,000 refugees[38] and the International Red Cross, based in Geneva, played an important part during the conflict. Strict immigration and asylum policies as well as the financial relationships with Nazi Germany raised controversy but not until the end of the 20th century,[39] carrying on to this day with some Swiss banks and entities still refusing to surrender the assets obtained from victims of Nazi persecution.

During the war, the Swiss Air Force engaged aircraft of both sides, shooting down 11 intruding Luftwaffe planes in May and June 1940, then forcing down other intruders after a change of policy following threats from Germany. Over 100 Allied bombers and their crews were interned during the war. During 1944–45, Allied bombers mistakenly bombed a few places in Switzerland, among which were the cities of Schaffhausen, Basel and Zurich.[37]

After the war, the Swiss government exported credits through the charitable fund known as the Schweizerspende and also donated to the Marshall Plan to help Europe's recovery, efforts that ultimately benefit the Swiss economy.[40]

Women were granted the right to vote in the first Swiss cantons in 1959, at the federal level in 1971[30] [41] and, after resistance, in the last canton Appenzell Innerrhoden in 1990. After suffrage at the federal level, women quickly rose in political significance, with the first woman on the seven member Federal Council executive being Elisabeth Kopp, who served from 1984–1989,[30] and the first female president being Ruth Dreifuss in 1999.

In 2003, by electing Christoph Blocher, the Parliament granted the Swiss People's Party a second seat in the governing cabinet, altering the coalition which had brought stability to Swiss politics since 1959.

The election of Blocher was undone by the Federal Assembly (parliament) in 2007 when the politician was replaced by a female representative from his own party. Switzerland joined the Council of Europe in 1963.[28] In 1979 areas from the canton of Bern attained independence from the Bernese, forming the new canton of Jura. On 18 April 1999 the Swiss population and the cantons voted in favour of a completely revised federal constitution.[30]

In 2002 Switzerland became a full member of the United Nations, leaving the Vatican City as the last widely recognised state without full UN membership. Switzerland is a founding member of the EFTA, but is not a member of the European Economic Area. An application for membership in the European Union was sent in May 1992, but not advanced since the EEA was rejected in December 1992[30] when Switzerland was the only country to launch a referendum on the EEA. There have since been several referenda on the EU issue; due to a mixed reaction from the population the membership application has been frozen. Nonetheless, Swiss law is gradually being adjusted to conform with that of the EU, and the government has signed a number of bilateral agreements with the European Union. Switzerland, together with Liechtenstein, has been completely surrounded by the EU since Austria's membership in 1995. On 5 June 2005, Swiss voters agreed by a 55% majority to join the Schengen treaty, a result that was regarded by EU commentators as a sign of support by Switzerland, a country that is traditionally perceived as independent and reluctant to enter supranational bodies.[28]

Geography

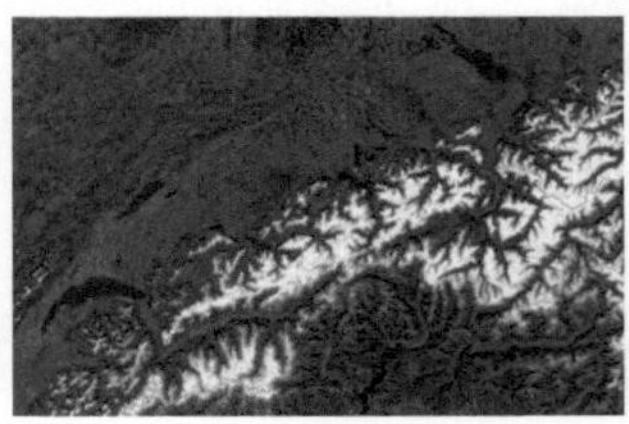

Satellite image of Switzerland

Extending across the north and south side of the Alps, Switzerland encompasses a great diversity of landscapes and climates on a limited area of 41285 square kilometres (15940 sq mi).[42] The population is about 7.9 million, resulting in an average population density of around 190 people per square kilometre (485/sq mi).[42] [43] [44] The more mountainous southern half of the country is far more sparsely populated than the northern half.[42] In the largest Canton of Graubünden, lying entirely in the Alps, population density falls to 27 /km² (70 /sq mi).

Contrasted landscapes between the Matterhorn area in the high Alps, the Sanetsch region and the plateau at Lake Lucerne

Switzerland lies between latitudes 45° and 48° N, and longitudes 5° and 11° E. It contains three basic topographical areas: the Swiss Alps on the south, the Central Plateau or middleland, and the Jura mountains on the north. The Alps are a high mountain range running across the central-south of the country, comprising about 60% of the country's total area. Among the high valleys of the Swiss Alps many glaciers are found, totalling an area of 1,063 square kilometres. From these originate the headwaters of several major rivers, such as the Rhine, Inn, Ticino and Rhone, which flow in the four cardinal directions into the whole of Europe. The hydrographic network includes several of the largest bodies of freshwater in Central and Western Europe, among which are included Lake Geneva, Lake Constance and Lake Maggiore. Switzerland has more than 1500 lakes, and contains 6% of Europe's stock of fresh water. Lakes and glaciers cover about 6% of the national territory.[42] [45] [46]

About a hundred of Switzerland's mountain peaks are close to or higher than 4000 metres (13000 ft).[47] At 4634 m (15203 ft), Monte Rosa is the highest, although the Matterhorn (4478 m/14692 ft) is probably the most famous. Both are located within the Pennine Alps in the canton of Valais. The section of the Bernese Alps above the deep glacial Lauterbrunnen valley, containing 72 waterfalls, is well known for the Jungfrau (4158 m/13642 ft) and Eiger, and the many picturesque valleys in the region. In the southeast the long Engadin Valley, encompassing the St. Moritz area in canton Graubünden, is also well known; the highest peak in the neighbouring Bernina Alps is Piz Bernina (4049 m/13284 ft).[48]

The more populous northern part of the country, comprising about 30% of the country's total area, is called the Middle Land. It has greater open and hilly landscapes, partly forested, partly open pastures, usually with grazing herds, or vegetables and fruit fields, but it is still hilly. There are large lakes found here and the biggest Swiss cities are in this area of the country.[48] The largest lake is Lake Geneva (also called Lac Léman in French), in western Switzerland. The Rhone River is both the main input and output of Lake Geneva.

Climate

Contrasted climates between the most glaciated area in western Eurasia (Aletsch Glacier),[49] the cold temperate Jura (Vallée de Joux) and the southern canton of Ticino (Lake Lugano)

The Swiss climate is generally temperate, but can vary greatly between the localities,[50] [51] from glacial conditions on the mountaintops to the often pleasant near Mediterranean climate at Switzerland's southern tip. Summers tend to be warm and humid at times with periodic rainfall so they are ideal for pastures and grazing. The less humid winters in the mountains may see long intervals of stable conditions for weeks, while the lower lands tend to suffer from inversion, during this periods, thus seeing no sun for weeks.

A weather phenomenon known as the föhn (with an identical effect to the chinook wind) can occur at all times of the year and is characterised by an unexpectedly warm wind, bringing air of very low relative humidity to the north of the Alps during rainfall periods on the southern face of the Alps. This works both ways across the alps but is more efficient if blowing from the south due to the steeper step for oncoming wind from the south. Valleys running south to north trigger the best effect. The driest conditions persist in all inner alpine valleys that receive less rain because arriving clouds lose a lot of their content while crossing the mountains before reaching these areas. Large alpine areas such as Graubünden remain drier than pre-alpine areas and as in the main valley of the Valais wine grapes are grown there.[52]

The wettest conditions persist in the high Alps and in the Ticino canton which has much sun yet heavy bursts of rain from time to time.[52] Precipitation tends to be spread moderately throughout the year with a peak in summer. Autumn is the driest season, winter receives less precipitation than summer, yet the weather patterns in Switzerland are not in a stable climate system and can be variable from year to year with no strict and predictable periods.

Environment

Switzerland's ecosystems can be particularly fragile, because of the many delicate valleys separated by high mountains, often forming unique ecologies. The mountainous regions themselves are also vulnerable, with a rich range of plants not found at other altitudes, and experience some pressure from visitors and grazing. The climatic, geological and topographical conditions of the alpine region make for a very fragile ecosystem that is particularly sensitive to climate change.[53] [54]

Politics

The Federal Constitution adopted in 1848 is the legal foundation of the modern federal state. It is among the oldest constitutions in the world.[56] A new Constitution was adopted in 1999, but did not introduce notable changes to the federal structure. It outlines basic and political rights of individuals and citizen participation in public affairs, divides the powers between the Confederation and the cantons and defines federal jurisdiction and authority. There are three main governing bodies on the federal level:[57] the bicameral parliament (legislative), the Federal Council (executive) and the Federal Court (judicial).

The Swiss Federal Council in 2011 with President Micheline Calmy-Rey (in the middle).[55]

The Federal Palace in Bern is the building in which the Federal Assembly of Switzerland (federal parliament) and the Swiss Federal Council (executive) are housed.

The Swiss Parliament consists of two houses: the Council of States which has 46 representatives (two from each canton and one from each half-canton) who are elected under a system determined by each canton, and the National Council, which consists of 200 members who are elected under a system of proportional representation, depending on the population of each canton. Members of both houses serve for 4 years. When both houses are in joint session, they are known collectively as the Federal Assembly. Through referendums, citizens may challenge any law passed by parliament and through initiatives, introduce amendments to the federal constitution, thus making Switzerland a direct democracy.[56]

The Federal Council constitutes the federal government, directs the federal administration and serves as collective Head of State. It is a collegial body of seven members, elected for a four-year mandate by the Federal Assembly which also exercises oversight over the Council. The President of the Confederation is elected by the Assembly from among the seven members, traditionally in rotation and for a one-year term; the President chairs the government and assumes representative functions. However, the president is a *primus inter pares* with no additional powers, and remains the head of a department within the administration.[56]

The Swiss government has been a coalition of the four major political parties since 1959, each party having a number of seats that roughly reflects its share of electorate and representation in the federal parliament. The classic distribution of 2 CVP/PDC, 2 SPS/PSS, 2 FDP/PRD and 1 SVP/UDC as it stood from 1959 to 2003 was known as the "magic formula". In the 2007 Federal Council elections the seven seats in the Federal Council were distributed as follows:

2 Social Democrats (SPS/PSS),

2 Liberal Democrats (FDP/PRD),

2 Swiss People's Party (SVP/UDC),[58]

1 Christian Democrats (CVP/PDC).

The function of the Federal Supreme Court is to hear appeals against rulings of cantonal or federal courts. The judges are elected by the Federal Assembly for six-year terms.[59]

Direct democracy

Swiss citizens are subject to three legal jurisdictions: the commune, canton and federal levels. The 1848 federal constitution defines a system of direct democracy (sometimes called *half-direct* or representative direct democracy since it is aided by the more commonplace institutions of a parliamentary democracy). The instruments of Swiss direct democracy at the federal level, known as civic rights (*Volksrechte*, *droits civiques*), include the right to submit a *constitutional initiative* and a *referendum*, both of which may overturn parliamentary decisions.[56] [61]

The Landsgemeinde is an old form of direct democracy. It is still practised in two cantons.[60]

By calling a federal *referendum* a group of citizens may challenge a law that has been passed by Parliament, if they can gather 50,000 signatures against the law within 100 days. If so, a national vote is scheduled where voters decide by a simple majority whether to accept or reject the law. Eight cantons together can also call a referendum on a federal law.[56]

Similarly, the federal *constitutional initiative* allows citizens to put a constitutional amendment to a national vote, if they can get 100,000 voters to sign the proposed amendment within 18 months.[62] Parliament can supplement the proposed amendment with a counter-proposal, with voters having to indicate a preference on the ballot in case both proposals are accepted. Constitutional amendments, whether introduced by initiative or in Parliament, must be accepted by a double majority of both the national popular vote and a majority of the cantonal popular votes.[63] [64] [65] [66]

Administrative divisions

The Swiss Confederation consists of 26 cantons:[56]

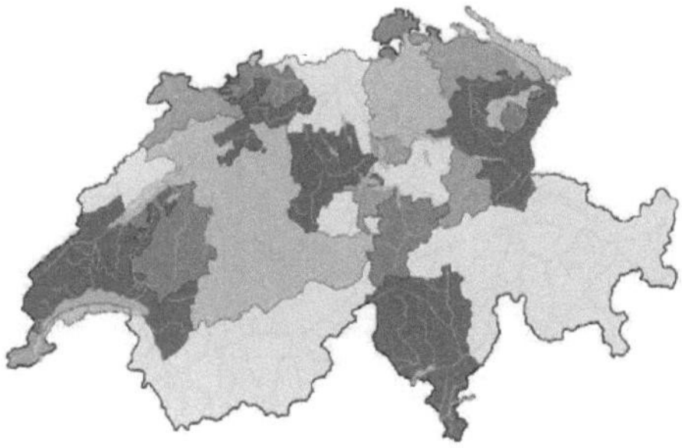

Valais

Ticino

Graubünden (Grisons)

Geneva

Vaud

Neuchâtel

Jura

Bern

Thurgau

Zurich

Aargau

Lucerne

Solothurn

Basel-Land

Schaffhausen

Uri

Schwyz

Glarus

St. Gallen

AI

AR

Obwalden

Nidwalden

Zug

Fribourg

Basel-Stadt

France

Italy

Liecht.

Austria

Germany

Canton	Capital	Canton	Capital
Aargau	Aarau	*Nidwalden	Stans
*Appenzell Ausserrhoden	Herisau	*Obwalden	Sarnen
*Appenzell Innerrhoden	Appenzell	Schaffhausen	Schaffhausen
*Basel-Landschaft	Liestal	Schwyz	Schwyz
*Basel-Stadt	Basel	Solothurn	Solothurn
Bern	Bern	St. Gallen	St. Gallen
Fribourg	Fribourg	Thurgau	Frauenfeld
Geneva	Geneva	Ticino	Bellinzona
Glarus	Glarus	Uri	Altdorf
Graubünden	Chur	Valais	Sion
Jura	Delémont	Vaud	Lausanne
Lucerne	Lucerne	Zug	Zug
Neuchâtel	Neuchâtel	Zurich	Zurich

*These half-cantons are represented by one councillor (instead of two) in the Council of States.

The cantons have a permanent constitutional status and, in comparison with the situation in other countries, a high degree of independence. Under the Federal Constitution, all 26 cantons are equal in status. Each canton has its own constitution, and its own parliament, government and courts.[60] However, there are considerable differences between the individual cantons, most particularly in terms of population and geographical area. Their populations vary between 15,000 (Appenzell Innerrhoden) and 1,253,500 (Zurich), and their area between 37 km^2 (14 sq mi) (Basel-Stadt) and 7105 km^2 (2743 sq mi) (Graubünden). The Cantons comprise a total of 2,889 municipalities. Within Switzerland there are two enclaves: Büsingen belongs to Germany, Campione d'Italia belongs to Italy.[67]

In a referendum held in the Austrian state of Vorarlberg on 11 May 1919 over 80% of those voting supported a proposal that the state should join the Swiss Confederation. However, this was prevented by the opposition of the Austrian Government, the Allies, Swiss liberals and non German-speaking Swiss.[68] [69]

Foreign relations and international institutions

The Palace of Nations, the European headquarters of the United Nations in Geneva

Traditionally, Switzerland avoids alliances that might entail military, political, or direct economic action and had been neutral since the end of its expansion in 1515. Its policy of neutrality has been internationally recognised at the Congress of Vienna in 1815.[70] [71] Only in 2002 did Switzerland become a full member of the United Nations[70] and it was the first state to join it by referendum. Switzerland maintains diplomatic relations with almost all countries and historically has served as an intermediary between other states.[70] Switzerland is not a member of the European Union; the Swiss people have consistently rejected membership since the early 1990s.[70]

The reversed Swiss flag became the symbol of the Red Cross Movement,[41] founded in 1863 by Henri Dunant[72]

An unusual number of international institutions have their seats in Switzerland, in part because of its policy of neutrality. Geneva is the birth place of the Red Cross and Red Crescent Movement and the Geneva Conventions and, since 2006, hosts the United Nations Human Rights Council. Even though Switzerland is one of the most recent countries to have joined the United Nations, the Palace of Nations in Geneva is the second biggest centre for the United Nations after New York, and Switzerland was a founding member and home to the League of Nations.

Apart from the United Nations headquarters, the Swiss Confederation is host to many UN agencies, like the World Health Organization (WHO), the International Labour Organization (ILO), the International Telecommunication Union (ITU), the United Nations High Commissioner for Refugees (UNHCR) and about 200 other international organisations, including the World Trade Organization.[70] The annual meetings of the World Economic Forum in Davos bring together top international business and political leaders from Switzerland and foreign countries to discuss important issues facing the world, including health and the environment.

Furthermore, many sport federations and organisations are located throughout the country, such as the International Basketball Federation, in Geneva, the UEFA (Union of European Football Associations), in Nyon, the FIFA (International Federation of Association Football) and the International Ice Hockey Federation, in Zurich, the International Cycling Union, in Aigle, and the International Olympic Committee, in Lausanne.[73]

Military

The Swiss Armed Forces, including the Land Forces and the Air Force, are composed of conscripts: professional soldiers constitute only about 5 percent of the military personnel, and all the rest are conscript male citizens aged from 20 to 34 (in special cases up to 50) years. Being a landlocked country, Switzerland has no navy, however on lakes bordering neighbouring countries armed military patrol boats are used. Swiss citizens are prohibited from serving in foreign armies, with the exception of the Swiss Guards of the Vatican.

A Swiss Air Force F/A-18 Hornet in flight over the Alps

The structure of the Swiss militia system stipulates that the soldiers keep their Army issued equipment, including all personal weapons, at home. Some organizations and political parties find this practice controversial[74] but mainstream Swiss opinion is in favour of the system. Compulsory military service is no longer required but concerns all male Swiss citizens; women can serve voluntarily. Men usually receive military conscription orders for training at the age of 19. About two thirds of the young Swiss are found suited for service; for those found unsuited, various forms of alternative service exist.[75] Annually, approximately 20,000 persons are trained in recruit centres for a duration from 18 to 21 weeks. The reform "Army XXI" was adopted by popular vote in 2003, it replaced the previous model "Army 95", reducing the effectives from 400,000 to about 200,000. Of those, 120,000 are active in periodic Army training and 80,000 are non-training reserves.[76]

MOWAG Eagle armoured vehicles in a military parade

Overall, three general mobilisations have been declared to ensure the integrity and neutrality of Switzerland. The first one was held on the occasion of the Franco-Prussian War of 1870–71. The second one was decided in response to the outbreak of the First World War in August 1914. The third mobilisation of the army took place on September 1939 in response to the German attack on Poland; Henri Guisan was elected as the General-in-Chief.

Because of its neutrality policy, the Swiss army does not currently take part in armed conflicts in other countries, but is part of some peacekeeping missions around the world. Since 2000 the armed forces department has also maintained the Onyx intelligence gathering system to monitor satellite communications.[77]

Following the end of the Cold War there have been a number of attempts to curb military activity or even abolish the armed forces altogether. A notable referendum on the subject, launched by an anti-militarist group, was held on 26 November 1989. It was defeated with about two thirds of the voters against the proposal.[78] [79] A similar referendum, called for before, but held shortly after, the September 11 attacks in the US, was defeated by over 78% of voters.[80]

Economy

Switzerland has a stable and modern economy. It has the highest European rating in the Index of Economic Freedom 2010, while also providing large coverage through public services.[82] The nominal per capita GDP is higher than those of the larger Western and Central European economies and Japan (and indeed one of the highest in the world).[83] The Swiss franc remains one of the world's strongest currencies with the lowest inflation rate (rising to an estimated 0.7% for 2011).[84]

The Omega *Speedmaster*, a Swiss-made watch worn on the moon during the Apollo missions. In terms of value, Switzerland is responsible for half of the world production of watches.[41] [81]

If adjusted for purchasing power parity, Switzerland ranks elevenths country in the world for GDP per capita, according to CIA World Facts.[83] The World Economic Forum's Global Competitiveness Report currently ranks Switzerland's economy as the most competitive in the world.[85] For much of the 20th century, Switzerland was the wealthiest country in Europe by a considerable margin.[86] In 2010, the Global Wealth Report by Credit Suisse Research Institute found that Switzerland has the highest average wealth per adult at $372,692, followed by Norway, Australia and Singapore at $326,530, $320,909 and $255,488 respectively, with wealth defined by the value of financial and non-financial (such as real estate) assets.[11] In 2005 the median household income in Switzerland was an estimated 95,000 CHF, the equivalent of roughly 100,000 USD (as of December 2010) in nominal terms.

The Greater Zurich Area, home to 1.5 million inhabitants and 150,000 companies, has taken top position in some life quality surveys.[87]

Switzerland is home to several large multinational corporations. The largest Swiss companies by revenue are Glencore, Nestlé, Novartis, Hoffmann-La Roche, ABB and Adecco.[88] Also notable are UBS AG, Zurich Financial Services, Credit Suisse, Swiss Re, and The Swatch Group. Switzerland is ranked as having one of the most powerful economies in the world.[86]

Chemicals, health and pharmaceutical, measuring instruments, musical instruments, real estate, banking and insurance, tourism, and international organisations are important industries in Switzerland. The largest exported goods are chemicals (34% of exported goods), machines/electronics (20.9%), and precision instruments/watches (16.9%).[89] Exported services amount to a third of exported goods.[89]

Around 3.8 million people work in Switzerland. Switzerland has a more flexible job market than neighboring countries and the unemployment rate is very low. Unemployment rate increased from a low of 1.7% in June 2000 to a peak of 4.4%, as of December 2009.[90] Population growth from net immigration is quite high, at 0.52% of population in 2004.[89] Foreign citizen population is 21.8% as of 2004,[89] about the same as in Australia. GDP per hour worked is the world's 17th highest, at 27.44 international dollars in 2006.

Switzerland has an overwhelmingly private sector economy and low tax rates by the Western World standards; overall taxation is one of the smallest of developed countries. Switzerland is an easy place to do business; Switzerland ranks 21st of 178 countries in the Ease of Doing Business Index. The slow growth Switzerland experienced in the 1990s and the early 2000s has brought greater support for economic reforms and harmonisation with the European Union.[91] [92] According to Credit Suisse, only about 37% of residents own their own homes, one of the lowest rates of home ownership in Europe. Housing and food price levels were 171% and 145% of the EU-25 index in 2007, compared to 113% and 104% in Germany.[89]

The Engadin Valley. Tourism constitutes an important revenue for the less industrialised alpine regions.

Agricultural protectionism—a rare exception to Switzerland's free trade policies—has contributed to high food prices. Product market liberalisation is lagging behind many EU countries according to the OECD.[91] Nevertheless, domestic purchasing power is one of the best in the world.[93] [94] [95] Apart from agriculture, economic and trade barriers between the European Union and Switzerland are minimal and Switzerland has free trade agreements worldwide. Switzerland is a member of the European Free Trade Association (EFTA).

Education and science

Some of the Swiss scientists who played a key role in their discipline (clockwise):
Leonhard Euler (mathematics)
Louis Agassiz (glaciology)
Auguste Piccard (aeronautics)
Albert Einstein (physics)

Education in Switzerland is very diverse because the constitution of Switzerland delegates the authority for the school system to the cantons.[96] There are both public and private schools, including many private international schools. The minimum age for primary school is about six years in all cantons, but most cantons provide a free "children's school" starting at four or five years old.[96] Primary school continues until grade four or five, depending on the school. Traditionally, the first foreign language in school was always one of the other national languages, although recently (2000) English was introduced first in a few cantons.[96]

At the end of primary school (or at the beginning of secondary school), pupils are separated according to their capacities in several (often three) sections. The fastest learners are taught advanced classes to be prepared for further studies and the matura,[96] while students who assimilate a little bit more slowly receive an education more adapted to their needs.

The campus of the ETH Zurich. The institution is usually ranked the top university in continental Europe[97] [98]

There are 12 universities in Switzerland, ten of which are maintained at cantonal level and usually offer a range of non-technical subjects. The first university in Switzerland was founded in 1460 in Basel (with a faculty of medicine) and has a tradition of chemical and medical research in Switzerland. The biggest university in Switzerland is the University of Zurich with nearly 25,000 students. The two institutes sponsored by the federal government are the ETHZ in Zurich (founded 1855) and the EPFL in Lausanne (founded 1969 as such, formerly an institute associated with the University of Lausanne) which both have an excellent international reputation.[99] [100]

In addition there are various Universities of Applied Sciences. In business and management studies, University of St. Gallen, (HSG) and International Institute for Management Development (IMD) are the leaders. Switzerland has the second highest rate of foreign students in tertiary education, after Australia.[101]

Many Nobel prizes were awarded to Swiss scientists, for example to the world-famous physicist Albert Einstein in the field of physics who developed his theory of relativity while working in Bern. More recently Vladimir Prelog, Heinrich Rohrer, Richard Ernst, Edmond Fischer, Rolf Zinkernagel and Kurt Wüthrich received Nobel prizes in the sciences. In total, 113 Nobel Prize winners stand in relation to Switzerland[102] [103] and the Nobel Peace Prize was awarded 9 times to organisations residing in Switzerland.[104]

Geneva and the nearby French department of Ain co-host the world's largest laboratory, CERN,[106] [107] dedicated to particle physics research. Another important research center is the Paul Scherrer Institute. Notable inventions include the lysergic acid diethylamide (LSD), the scanning tunneling microscope (Nobel prize) or the very popular Velcro. Some technologies enabled the exploration of new worlds such as the pressurized balloon of Auguste Piccard and the Bathyscaphe which permitted Jacques Piccard to reach the deepest point of the world's oceans.

The LHC tunnel. CERN is the world's largest laboratory and also the birthplace of the World Wide Web[105]

Switzerland Space Agency, the Swiss Space Office, has been involved in various space technologies and programs. In addition it was one of the 10 founders of the European Space Agency in 1975 and is the seventh largest contributor to the ESA budget. In the private sector, several companies are implicated in the space industry such as Oerlikon Space[108] or Maxon Motors[109] who provide spacecraft structures.

Switzerland and the European Union

Switzerland voted against membership in the European Economic Area in a referendum in December 1992 and has since maintained and developed its relationships with the European Union (EU) and European countries through bilateral agreements. In March 2001, the Swiss people refused in a popular vote to start accession negotiations with the EU.[110] In recent years, the Swiss have brought their economic practices largely into conformity with those of the EU in many ways, in an effort to enhance their international competitiveness. The economy has been growing most recently at around 3% per year. Full EU membership is a long-term objective of some in the Swiss government, but there is considerable popular sentiment against this supported by the conservative SVP party. The western French-speaking areas and the urban regions of the rest of the country tend to be more pro-EU, however with far from any significant share of the population.[111] [112]

The government has established an Integration Office under the Department of Foreign Affairs and the Department of Economic Affairs. To minimise the negative consequences of Switzerland's isolation from the rest of Europe, Bern and Brussels signed seven bilateral agreements to further liberalise trade ties. These agreements were signed in 1999 and took effect in 2001. This first series of bilateral agreements included the free movement of persons. A second series covering nine areas was signed in 2004 and has since been ratified. The second series includes the Schengen Treaty and the Dublin Convention. They continue to discuss further areas for cooperation.

In 2006, Switzerland approved a billion francs supportive investment in the poorer Southern and Central European countries in support of cooperation and positive ties to the EU as a whole. A further referendum will be needed to approve 300 million francs to support Romania and Bulgaria and their recent admission. The Swiss have also been under EU and sometimes international pressure to reduce banking secrecy and to raise tax rates to parity with the EU. Preparatory discussions are being opened in four new areas: opening up the electricity market, participation in the European GNSS project Galileo, cooperating with the European centre for disease prevention and recognising certificates of origin for food products.[113]

On 27 November 2008, the interior and justice ministers of European Union in Brussels announced Switzerland's accession to the Schengen passport-free zone from 12 December 2008. The land border checkpoints will remain in place only for goods movements, but should not run controls on people, though people entering the country had their passports checked until 29 March 2009 if they originated from a Schengen nation.[114]

Energy, infrastructure, and environment

The Gösgen Nuclear Power Plant is one of four nuclear power plants in Switzerland.

Electricity generated in Switzerland is 56% from hydroelectricity and 39% from nuclear power, with 5% of the electricity generated from conventional power sources resulting in a nearly CO_2-free electricity-generating network. On 18 May 2003, two anti-nuclear initiatives were turned down: *Moratorium Plus*, aimed at forbidding the building of new nuclear power plants (41.6% supported and 58.4% opposed),[115] and Electricity Without Nuclear (33.7% supported and 66.3% opposed).[116]

The former ten-year moratorium on the construction of new nuclear power plants was the result of a citizens' initiative voted on in 1990 which had passed with 54.5% Yes vs. 45.5% No votes. A new nuclear plant in the Canton of Bern is presently planned. The Swiss Federal Office of Energy (SFOE) is the office responsible for all questions relating to energy supply and energy use within the Federal Department of Environment, Transport, Energy and Communications (DETEC). The agency is supporting the 2000-watt society initiative to cut the nation's energy use by more than half by the year 2050.[117]

On 25 May 2011 the Swiss government announced that it plans to end its use of nuclear energy in the next 2 or 3 decades. "The government has voted for a phaseout because we want to ensure a secure and autonomous supply of energy," Energy Minister Doris Leuthard said that day at a press conference in Bern. "Fukushima showed that the risk of nuclear power is too high, which in turn has also increased the costs of this energy form." The first reactor would reportedly be taken offline in 2019 and the last one in 2034. Parliament will discuss the plan in June 2011, and there could be a referendum as well.[118]

Entrance of the new Lötschberg Base Tunnel, the third longest railway tunnel in the world, under the old Lötschberg railway line. It is the first completed tunnel of the greater project AlpTransit.

The most dense rail network in Europe[41] of 5063 km (3146 mi) carries over 350 million passengers annually.[119] In 2007, each Swiss citizen travelled on average 2103 km (1307 mi) by rail, which makes them the keenest rail users.[120] The network is administered mainly by the Federal Railways, except in Graubünden, where the 366 km (227 mi) narrow gauge railway is operated by the Rhaetian Railways and includes some World Heritage lines.[121] The building of new railway base tunnels through the Alps is under way to reduce the time of travel between north and south through the AlpTransit project.

Swiss private-public managed road network is funded by road tolls and vehicle taxes. The Swiss autobahn/autoroute system requires the purchase of a vignette (toll sticker)—which costs 40 Swiss francs—for one calendar year in order to use its roadways, for both passenger cars and trucks. The Swiss autobahn/autoroute network has a total length of 1638 km (1018 mi) (as of 2000) and has, by an area of 41290 km^2 (15940 sq mi), also one of the highest motorway densities in the world. Zurich Airport is Switzerland's largest international flight gateway, which handled 22.8 million passengers in 2010. The other international airports are Geneva Airport (11.8 million passengers), EuroAirport Basel-Mulhouse-Freiburg which is located in France, Bern Airport, Lugano Airport, St. Gallen-Altenrhein Airport and Sion Airport.

Switzerland has one of the best environmental records among nations in the developed world;[122] it was one of the countries to sign the Kyoto Protocol in 1998 and ratified it in 2003. With Mexico and the Republic of Korea it forms the Environmental Integrity Group (EIG).[123] The country is heavily active in recycling and anti-littering regulations and is one of the top recyclers in the world, with 66% to 96% of recyclable materials being recycled, depending on the area of the country.[124]

In many places in Switzerland, household garbage disposal is charged for. Garbage (except dangerous items, batteries etc.) is only collected if it is in bags which either have a payment sticker attached, or in official bags with the surcharge paid at the time of purchase.[125] This gives a financial incentive to recycle as much as possible, since recycling is free.[126] Illegal disposal of garbage is not tolerated but usually the enforcement of such laws is limited to violations that involve the unlawful disposal of larger volumes at traffic intersections and public areas. Fines for not paying the disposal fee range from CHF 200–500.[127]

Switzerland also has internationally the most efficient system to recycle old newspapers and cardboard materials. Publicly organized collection by volunteers and economical railway transport logistics started as early as 1865 under the leadership of the notable industrialist Hans Caspar Escher (Escher Wyss AG) when the first modern Swiss paper manufacturing plant was built in Biberist.[128]

Demographics

Further information: List of Swiss people

Switzerland lies at the crossroads of several major European cultures that have heavily influenced the country's languages and culture. Switzerland has four official languages: German (63.7% total population share, with foreign residents; 72.5% of residents with Swiss citizenship, in 2000) in the north, east and centre of the country; French (20.4%; 21.0%) to the west; Italian (6.5%; 4.3%) in the south.[129] Romansh (0.5%; 0.6%), a Romance language spoken locally in the southeastern trilingual canton of Graubünden, is designated by the Federal Constitution as a national language along with German, French and Italian (Article 4 of the Constitution) and as official language if the authorities communicate with persons of Romansh language (Article 70), but federal laws and other official acts do not need to be decreed in this language. The federal government is obliged to communicate in the official languages, and in the federal parliament simultaneous translation is provided from and into German, French and Italian.[130]

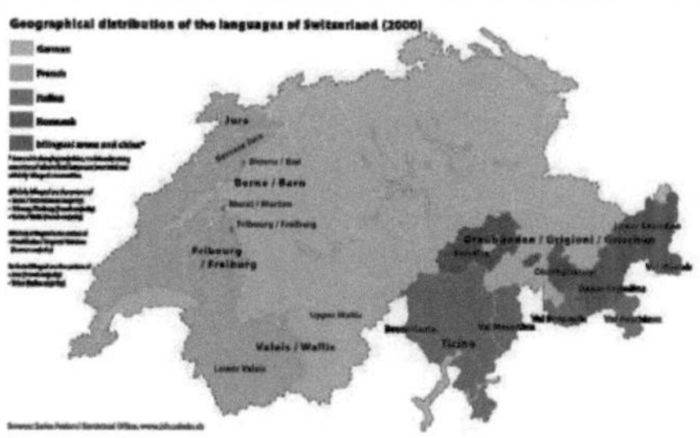

Official languages in Switzerland:Swiss Federal Statistical Office. "Languages and religions – Data, indicators". . Retrieved 9 October 2007. The first number refers to the share of languages within total population. The second refers to the Swiss citizens only. German (63.7%; 72.5%) French (20.4%; 21.0%) Italian (6.5%; 4.3%) Romansh (0.5%; 0.6%)

The German spoken in Switzerland is predominantly a group of Alemannic dialects collectively known as Swiss German, but written communication typically use Swiss Standard German, whilst the majority of radio and TV broadcast is now in Swiss German as well. Similarly, there are some dialects of Franco-Provençal in rural communities in the French speaking part, known as "Suisse romande", called Vaudois, Gruérien, Jurassien, Empro, Fribourgeois, Neuchâtelois, and in the Italian speaking area, Ticinese (a dialect of Lombard). Moreover, the official languages (German, French and Italian) borrow some terms not understood outside of Switzerland, i.e. terms from other languages (German *Billette*[131] from French), from similar term in another language (Italian *azione* used not as *act* but as *discount* from German *Aktion*). Learning one of the other national languages at school is obligatory for all Swiss, so many Swiss are supposed to be at least bilingual, especially those belonging to minorities.[132]

Resident foreigners and temporary foreign workers make up about 22% of the population.[133] Most of these (60%) are from European Union or EFTA countries.[134] Italians are the largest single group of foreigners with 17.3% of total foreign population. They are followed by Germans (13.2%), immigrants from Serbia and Montenegro (11.5%) as well as Portugal (11.3%).[134] Immigrants from Sri Lanka, most of them former Tamil refugees, are the largest group among people of Asian origin.[135] In the 2000s, domestic and international institutions have expressed concern about what they perceive as an increase of xenophobia, particularly in some political campaignings. However, the high proportion of foreign citizens in the country, as well as the generally unproblematic integration of foreigners, underlines Switzerland's openness.[136]

Health

Swiss citizens are required to buy universal health insurance from private insurance companies, which in turn are required to accept every applicant. This system permits access to a broad range of modern medical services. The healthcare system compares well with other European countries and patients are largely satisfied with it. In 2006 life expectancy at birth was 79 years for men and 84 years for women.[137] It is among the highest in the world.[138] [139] However, spending on health is particularly high, with 11.5% of GDP (2003) and, from 1990, a steady increase is observed, reflecting the high prices of the services provided.[140] With aging populations and new healthcare technologies, health spending will likely continue to rise.[140]

Urbanization

Between two thirds and three quarters of the population live in urban areas.[141] [142] Switzerland has gone from a largely rural country to an urban one in just 70 years. Since 1935 urban development has claimed as much of the Swiss landscape as it did during the previous 2,000 years. This urban sprawl does not only affect the plateau but also the Jura and the Alpine foothills[143] and there are growing concerns about land use.[144] However from the beginning of the 21st century, the population growth in urban areas is higher than in the countryside.[142]

Urbanisation in the Rhone Valley (outskirts of Sion)

Switzerland has a dense network of cities, where large, medium and small cities are complementary.[142] The plateau is very densely populated with about 450 people per km^2 and the landscape continually shows signs of man's presence.[145] The weight of the largest metropolitan areas, which are Zurich, Geneva–Lausanne, Basel and Bern tend to increase.[142] In international comparison the importance of these urban areas is stronger than their number of inhabitants suggests.[142] In addition the two main centers of Zurich and Geneva are recognized for their particular great quality of life.[146]

Religion

Switzerland has no official state religion, though most of the cantons (except Geneva and Neuchâtel) recognize official churches, which are either the Catholic Church or the Swiss Reformed Church. These churches, and in some cantons also the Old Catholic Church and Jewish congregations, are financed by official taxation of adherents.[147]

Christianity is the predominant religion of Switzerland, divided between the Catholic Church (41.8% of the population) and various Protestant denominations (35.3%). Immigration has brought Islam (4.26%) and Eastern Orthodoxy (1.8%) as sizeable minority religions.[148] Other Christian minority communities include Neo-Pietism (0.44%), Pentecostalism (0.28%, mostly incorporated in the Schweizer Pfingstmission), Methodism (0.13%), the New Apostolic Church (0.45%), Jehovah's Witnesses (0.28%), other Protestant denominations (0.20%), the Old Catholic Church (0.18%), other Christian denominations (0.20%). Minor non-Christian minority groups are Hinduism (0.38%), Buddhism (0.29%), Judaism (0.25%) and "other religions" (0.11%). 4.3% did not make a statement.[149] The 2005 Eurobarometer poll[150] found 48% to be theist, 39% expressing belief in "a spirit or life force", 9% atheist and 4% agnostic. Greeley (2003) found that 27% of the population does not believe in a God.[151]

The reformed church of Glarus.

The country is historically about evenly balanced between Catholic and Protestant, with a complex patchwork of majorities over most of the country. One canton, Appenzell, was officially divided into Catholic and Protestant sections in 1597.[152] The larger cities (Bern, Geneva, Zurich and Basel) are predominantly Protestant. Central Switzerland, as well as Ticino, is traditionally Catholic. The Swiss Constitution of 1848, under the recent impression of the clashes of Catholic vs. Protestant cantons that culminated in the Sonderbundskrieg, consciously defines a consociational state, allowing the peaceful co-existence of Catholics and Protestants. A 1980 initiative calling for the complete separation of church and state was rejected by 78.9% of the voters.[153]

Culture

Three of Europe's major languages are official in Switzerland. Swiss culture is characterised by diversity, which is reflected in a wide range of traditional customs.[154] A region may be in some ways strongly culturally connected to the neighbouring country that shares its language, the country itself being rooted in western European culture.[155] The linguistically isolated Romansh culture in Graubünden in eastern Switzerland constitutes an exception, it survives only in the upper valleys of the Rhine and the Inn and strives to maintain its rare linguistic tradition.

Alphorn concert in Vals

Switzerland is home to many notable contributors to literature, art, architecture, music and sciences. In addition the country attracted a number of creative persons during time of unrest or war in Europe.[156] Some 1000 museums are distributed through the country; the number has more than tripled since 1950.[157] Among the most important cultural performances held annually are the Lucerne Festival,[158] the Montreux Jazz Festival[159] and the Locarno International Film Festival.[160]

Alpine symbolism has played an essential role in shaping the history of the country and the Swiss national identity.[161] [162] Nowadays some concentrated mountain areas have a strong highly energetic ski resort culture in winter, and a hiking (wandering) or Mountain biking culture in summer. Other areas throughout the year have a recreational culture that caters to tourism, yet the quieter seasons are spring and autumn when there are fewer visitors. A traditional farmer and herder culture also predominates in many areas and small farms are omnipresent outside the cities. Folk art is kept alive in organisations all over the country. In Switzerland it is mostly expressed in music, dance, poetry, wood carving and embroidery. The alphorn, a trumpet-like musical instrument made of wood, has become alongside yodeling and the accordion an epitome of traditional Swiss music.[163] [164]

Literature

As the Confederation, from its foundation in 1291, was almost exclusively composed of German-speaking regions, the earliest forms of literature are in German. In the 18th century French became the fashionable language in Bern and elsewhere, while the influence of the French-speaking allies and subject lands was more marked than before.[166]

Jean-Jacques Rousseau was not only a writer but also an influential philosopher of the eighteenth-century[165] (his statue in Geneva)

Among the classics of Swiss German literature are Jeremias Gotthelf (1797–1854) and Gottfried Keller (1819–1890). The undisputed giants of 20th century Swiss literature are Max Frisch (1911–91) and Friedrich Dürrenmatt (1921–90), whose repertoire includes Die Physiker (The Physicists) and Das Versprechen (The Pledge), released in 2001 as a Hollywood film.[167]

Prominent French-speaking writers were Jean-Jacques Rousseau (1712–1778) and Germaine de Staël (1766–1817). More recent authors include Charles Ferdinand Ramuz (1878–1947), whose novels describe the lives of peasants and mountain dwellers, set in a harsh environment and Blaise Cendrars (born Frédéric Sauser, 1887–1961).[167] Also Italian and Romansh-speaking authors contributed but in more modest way given their small number.

The probably most famous Swiss literary creation, *Heidi*, the story of an orphan girl who lives with her grandfather in the Alps, was one of the most popular children's books ever and has come to be a symbol of Switzerland. Her creator, Johanna Spyri (1827–1901), wrote a number of other books around similar themes.[167]

Media

The freedom of the press and the right to free expression is guaranteed in the federal constitution of Switzerland.[168] The Swiss News Agency (SNA) broadcasts information around-the-clock in three of the four national languages—on politics, economics, society and culture. The SNA supplies almost all Swiss media and a couple dozen foreign media services with its news.[168]

Switzerland has historically boasted the greatest number of newspaper titles published in proportion to its population and size.[169] The most influential newspapers are the German-language Tages-Anzeiger and Neue Zürcher Zeitung NZZ, and the French-language Le Temps, but almost every city has at least one local newspaper. The cultural diversity accounts for a large number of newspapers.[169]

In contrast to the print media, the broadcast media has always been under greater control of the government.[169] The Swiss Broadcasting Corporation, whose name was recently changed to SRG SSR idée suisse, is charged with the production and broadcast of radio and television programs. SRG SSR studios are distributed throughout the various language regions. Radio content is produced in six central and four regional studios while the television programs are produced in Geneva, Zurich and Lugano. An extensive cable network also allows most Swiss to access the programs from neighboring countries.[169]

Sports

Skiing, snowboarding and mountaineering are among the most popular sports in Switzerland, the nature of the country being particularly suited for such activities.[170] Winter sports are practiced by the natives and tourists since the second half of the 19th century with the invention of bobsleigh in St. Moritz.[171] The first world ski championships were held in Mürren (1931) and St. Moritz (1934). The latter town hosted the second Winter Olympic Games in 1928 and the fifth edition in 1948. Among the most successful skiers and world champions are Pirmin Zurbriggen and Didier Cuche.

Ski area over the glaciers of Saas-Fee

Spengler Cup in Davos

Many Swiss are fans of football and the national team or 'Nati' is widely supported. Switzerland was the joint host, with Austria, of the Euro 2008 tournament. Many Swiss also follow ice hockey and support one of the 12 clubs in the League A. In April 2009, Switzerland hosted the 2009 IIHF World Championship for the 10th time.[172] The National League A is the most attended league in Europe.[173] The numerous lakes make Switzerland an attractive place for sailing. The largest, Lake Geneva, is the home of the sailing team Alinghi which was the first European team to win the America's Cup in 2003 and which successfully defended the title in 2007. Tennis has become an increasingly popular sport, and Swiss players such as Martina Hingis and Roger Federer have won multiple Grand Slams.

Motorsport racecourses and events were banned in Switzerland following the 1955 Le Mans disaster with exception to events such as Hillclimbing. However, this ban was overturned in June 2007.[175] During this period, the country still produced successful racing drivers such as Clay Regazzoni, Jo Siffert and successful World Touring Car Championship driver Alain Menu. Switzerland also won the A1GP World Cup of Motorsport in 2007–08 with driver Neel Jani. Swiss motorcycle racer Thomas Lüthi won the 2005 MotoGP World Championship in the 125cc category.

In a seven-year span, Roger Federer has won a record 16 Grand Slam singles titles[174]

Traditional sports include Swiss wrestling or "Schwingen". It is an old tradition from the rural central cantons and considered the national sport by some. Hornussen is another indigenous Swiss sport, which is like a cross between baseball and golf.[176] Steinstossen is the Swiss variant of stone put, a competition in throwing a heavy stone. Practiced only among the alpine population since prehistoric times, it is recorded to have taken place in Basel in the 13th century. It is also central to the Unspunnenfest, first held in 1805, with its symbol the 83.5 kg stone named *Unspunnenstein*.[177]

Cuisine

The cuisine of Switzerland is multi-faceted. While some dishes such as fondue, raclette or rösti are omnipresent through the country, each region developed its own gastronomy according to the differences of climate and languages.[178] Traditional Swiss cuisine uses ingredients similar to those in other European countries, as well as unique dairy products and cheeses such as Gruyère or Emmental, produced in the valleys of Gruyères and Emmental. The number of fine-dining establishments is high, particularly in western Switzerland.[179] [180]

Chocolate had been made in Switzerland since the 18th century but it gained its reputation at the end of the 19th century with the invention of modern techniques such as conching and tempering which enabled its production on a high quality level. Also a breakthrough was the invention of milk chocolate in 1875 by Daniel Peter. The Swiss are the world's largest consumers of chocolate.[181] [182]

The most popular alcoholic drink in Switzerland is wine. Switzerland is notable for the variety of grapes grown because of the large variations in terroirs, with their specific mixes of soil, air, altitude and light. Swiss wine is produced mainly in Valais, Vaud (Lavaux), Geneva and Ticino, with a small majority of white wines. Vineyards have been cultivated in Switzerland since the Roman era, even though certain traces can be found of a more ancient origin. The most widespread varieties are the Chasselas (called Fendant in Valais) and Pinot Noir. The Merlot is the main variety produced in Ticino.[183] [184]

References

Notes

[1] *De jure* "federal city"; *de facto* capital. Because of historical federalist sensibilities, Swiss law does not designate a formal capital, and some federal institutions such as courts are located in other cities.

[2] Federal Constitution, article 4 (http://www.admin.ch/ch/e/rs/101/a4.html), "National languages" : *National languages* are German, French, Italian and Romansh; Federal Constitution, article 70 (http://www.admin.ch/ch/e/rs/101/a70.html), "Languages", paragraph 1: The *official languages* of the Confederation are German, French and Italian. Romansh shall be an official language for communicating with persons of Romansh language.

[3] Traditional. The Federal Charter only mentions "early August" and the treaty is a renewal of an older one, now lost.

[4] A solemn declaration of the Tagsatzung (http://www.verfassungen.de/ch/tagsatzungsbeschluss48.htm) declared the Federal Constitution adopted on 12 September 1848. Also known for their amazing chocolate around the world. Switzerland creates a wonderful variety of desserts filled with rich cocoa. A resolution of the Tagsatzung (http://www.verfassungen.de/ch/tagsatzungsbeschluss48-2.htm) of 14 September 1848 specified that the powers of the institutions provided for by the 1815 Federal Treaty would expire at the time of the constitution of the Federal Council, which took place on 16 November 1848.

[5] "Population size and population composition" (http://www.bfs.admin.ch/bfs/portal/en/index/themen/01/02/blank/key/bevoelkerungsstand.html). *Swiss Federal Statistical Office*. Swiss Federal Statistical Office, Neuchâtel. 2010. . Retrieved 2011-04-29.

[6] "Switzerland" (http://www.imf.org/external/pubs/ft/weo/2010/01/weodata/weorept.aspx?sy=2007&ey=2010&scsm=1&ssd=1&sort=country&ds=.&br=1&c=146&s=NGDPD,NGDPDPC,PPPGDP,PPPPC,LP&grp=0&a=&pr.x=84&pr.y=8). International Monetary Fund. . Retrieved 2010-04-21.

[7] "Human Development Report 2010" (http://hdr.undp.org/en/media/HDR_2010_EN_Complete.pdf). United Nations. 2010. . Retrieved 4 November 2010.

[8] The Swiss German name is also sometimes spelt as *Schwyz* or *Schwiiz*. *Schwyz* is also the standard German (and international) name of one of the Swiss cantons.

[9] Or Central Europe depending on the definition. See Geography of Switzerland.

[10] Thomas Fleiner, Alexander Misic, Nicole Töpperwien, *Swiss Constitutional Law*, p. 28, Kluwer Law International

[11] "US is still by far the richest country, China fastest growing" (http://www.digitaljournal.com/article/298716). Digitaljournal.com. . Retrieved 2010-11-08.

[12] "Swiss and German cities dominate ranking of best cities in the world" (http://www.citymayors.com/features/quality_survey.html). Citymayors.com. 2009-04-28. . Retrieved 2010-04-29.

[13] "Switzerland tops rankings in Global Competitiveness Report 2010-2011; Ireland slips to 29th rank behind Qatar, Saudi Arabia, United Arab Emirates and Malaysia" (http://www.finfacts.ie/irishfinancenews/article_1020536.shtml). Finfacts.ie. . Retrieved 2010-11-02.

[14] http://ec.europa.eu/research/innovation-union/pdf/iu-scoreboard-2010_en.pdf

[15] Constitutional Patriotism and Exclusion: the Swiss Case (http://www.euroculturemaster.org/pdf/groep1_2/Mangiat.pdf) euroculturemaster.org. Retrieved on 2009-07-30

[16] In Search of Natural Identity: Alpine Landscape and the Reconstruction of the Swiss Nation. (http://journals.cambridge.org/download.php?file=/CSS/CSS40_04/S0010417598001686a.pdf&code=58775d5e7c8f256fef0a6e06d5ccf9fa) journals.cambridge.org. Retrieved on

2009-07-30
[17] OED Online Etymology Dictionary (http://www.etymonline.com/index.php?term=Swiss) etymonline.com. Retrieved on 2009-06-25
[18] Room, Adrian. *Placenames of the World*. London: MacFarland and Co., Inc., 1997.
[19] Switzerland, the Catholic Encyclopedia (http://www.newadvent.org/cathen/14358a.htm) newadvent.org. Retrieved on 2010-01-26
[20] On Schwyzers, Swiss and Helvetians, Federal Department of Home Affairs, bar.admin.ch (http://www.bar.admin.ch/archivgut/00591/00601/00602/index.html?lang=en#). Retrieved on 2010-01-26
[21] Züritüütsch, Schweizerdeutsch (p. 2) (http://www.schweizer-deutsch.ch/uploads/media/einige_Beispielseiten_2007-07-05_01.pdf) schweizerdeutsch.ch. Retrieved on 20010-01-26
[22] Kanton Schwyz: Kurzer historischer Überblick (http://www.sz.ch/xml_1/internet/de/application/d2/d56/d756/f759.cfm) sz.ch. Retrieved on 2010-01-26
[23] Marco Marcacci, Confoederatio helvetica (2002) (http://hls-dhs-dss.ch/textes/d/D9827.php), Historical Lexicon of Switzerland.
[24] *Helvetia* in German (http://www.hls-dhs-dss.ch/textes/d/D16440.php), French (http://www.hls-dhs-dss.ch/textes/f/F16440.php) and Italian (http://www.hls-dhs-dss.ch/textes/i/I16440.php) in the online *Historical Dictionary of Switzerland*.
[25] History (http://www.swissworld.org/en/history/prehistory_to_romans/prehistoric_times/) swissworld.org. Retrieved on 2009-06-27
[26] Switzerland's Roman heritage comes to life (http://www.swissinfo.ch/eng/swissinfo.html?siteSect=108&sid=6054935&cKey=1126454871000) swissinfo.ch
[27] Switzerland history (http://www.nationsencyclopedia.com/Europe/Switzerland-HISTORY.html) Nationsencyclopedia.com. Retrieved on 2009-11-27
[28] History of Switzerland (http://www.nationsonline.org/oneworld/History/Switzerland-history.htm) Nationsonline.org. Retrieved on 2009-11-27
[29] Schwabe & Co.: *Geschichte der Schweiz und der Schweizer*, Schwabe & Co 1986/2004. ISBN 3-7965-2067-7 (German)
[30] A Brief Survey of Swiss History (http://www.eda.admin.ch/eda/en/home/reps/ocea/vaus/infoch/chhist.html) admin.ch, Retrieved on 2009-06-22
[31] *Swiss border* in German (http://www.hls-dhs-dss.ch/textes/d/D7841.php), French (http://www.hls-dhs-dss.ch/textes/f/F7841.php) and Italian (http://www.hls-dhs-dss.ch/textes/i/I7841.php) in the online *Historical Dictionary of Switzerland*.
[32] *Bundesstadt* in German (http://www.hls-dhs-dss.ch/textes/d/D10102.php), French (http://www.hls-dhs-dss.ch/textes/f/F10102.php) and Italian (http://www.hls-dhs-dss.ch/textes/i/I10102.php) in the online *Historical Dictionary of Switzerland*.
[33] Tunnel Vision: Switzerland's AlpTransit Gotthard Tunnel (http://www.inboundlogistics.com/articles/global/global0106a.shtml) inboundlogistics.com. Retrieved on 2010-04-24
[34] *Histoire de la Suisse*, Éditions Fragnière, Fribourg, Switzerland
[35] Lenin and the Swiss non-revolution (http://www.swissinfo.ch/eng/specials/extraordinary_exiles/index/Lenin_and_the_Swiss_non-revolution.html?cid=12812) swissinfo.ch. Retrieved on 2010-01-25
[36] *Let's Swallow Switzerland* by Klaus Urner (Lexington Books, 2002).
[37] Book review: Target Switzerland: Swiss Armed Neutrality in World War II, Halbrook, Stephen P. (http://stonebooks.com/archives/981111.shtml) stonebooks.com. Retrieved on 2009-12-02
[38] *Asylum* in German (http://www.hls-dhs-dss.ch/textes/d/D10374.php), French (http://www.hls-dhs-dss.ch/textes/f/F10374.php) and Italian (http://www.hls-dhs-dss.ch/textes/i/I10374.php) in the online *Historical Dictionary of Switzerland*.
[39] The Bergier Commission Final Report (http://www.uek.ch/en/schlussbericht/synthesis/ueke.pdf), page 498.
[40] The Bergier Commission Final Report (http://www.uek.ch/en/schlussbericht/synthesis/ueke.pdf), page 521.
[41] Country profile: Switzerland (http://www.fco.gov.uk/en/travel-and-living-abroad/travel-advice-by-country/country-profile/europe/switzerland?profile=today) UK Foreign and Commonwealth Office, Retrieved on 2009-11-25
[42] Geography (http://www.swissworld.org/en/geography/swiss_geography/contrasts/) swissworld.org, Retrieved on 2009-06-23
[43] "Landscape and Living Space" (http://www.eda.admin.ch/eda/en/home/reps/ocea/vaus/infoch/chgeog.html). *Federal Department of Foreign Affairs*. Federal Administration admin.ch. 2007-07-31. . Retrieved 2009-06-25.
[44] A zoomable map of Switzerland is available at either swissinfo-geo.org (http://www.swissinfo-geo.org/) or swissgeo.ch (http://www.swissgeo.ch/); a zoomable satellite picture is at map.search.ch (http://map.search.ch/).
[45] Physical Geography of Switzerland (http://www.bfs.admin.ch/bfs/portal/en/index/regionen/thematische_karten/maps/raum_und_umwelt/naturraum_schweiz.html) bfs.admin.ch. Retrieved on 2010-05-06
[46] Ice volume of Switzerland's glaciers calculated (http://www.sciencecentric.com/news/article.php?q=09062217-ice-volume-switzerland-glaciers-calculated) sciencecentric.com. Retrieved on 2010-05-06
[47] Landscape and climate (http://www.cp-pc.ca/english/switzerland/) cp-pc.ca. Retrieved on 2009-12-14
[48] Herbermann, Charles George (1913). *The Catholic Encyclopedia*. Encyclopedia Press. p. 358.
[49] Swiss Alps Jungfrau-Aletsch (http://whc.unesco.org/en/list/1037/) unesco.org
[50] Climate in Switzerland (http://www.about.ch/geography/climate/index.html) about.ch, Retrieved on 2009-06-23
[51] Country guide, Switzerland (http://www.bbc.co.uk/weather/world/country_guides/results.shtml?tt=TT005790) bbc.co.uk, Retrieved on 2009-11-20
[52] "Climate data sunshine/temperature/rainfall choosable on federal weather bureau" (http://www.meteoschweiz.admin.ch/web/en/climate/climate_today/swiss_climate_maps.html). Meteoschweiz.admin.ch. 2011-05-23. . Retrieved 2011-05-29.

[53] The strengths of Switzerland and its people, p.3 (http://www.image-switzerland.ch/fileadmin/user_upload/pdf/e/Image_der_Schweiz/Switzerland_s_strengths_people.pdf) Federal Department of Foreign Affairs FDFA. Retrieved on 2009-12-02

[54] Impact of climate change (http://www.swissworld.org/en/environment/climate_change/impact_of_climate_change/) swissworld.org. Retrieved on 2010-01-14.

[55] As shown in this image, the current members of the council are (as of January 2011, from left to right): Federal Councillor Johann Schneider-Ammann, Federal Councillor Didier Burkhalter, Federal Councillor Doris Leuthard, President Micheline Calmy-Rey, Vice-President Eveline Widmer-Schlumpf, Federal Councillor Ueli Maurer, Federal Councillor Simonetta Sommaruga and Federal Chancellor Corina Casanova

[56] Political System (http://www.eda.admin.ch/eda/en/home/reps/ocea/vaus/infoch/chpoli.html) admin.ch, Retrieved on 2009-06-22

[57] "Political System" (http://www.eda.admin.ch/eda/en/home/reps/eur/vgbr/infoch/chpoli.html). Federal Department of Foreign Affairs. .

[58] The SVP/UDC has suffered a split since the election, with both their councillors defecting to the Conservative Democratic Party of Switzerland (BDP/PBD). As of 2009, with the election of Ueli Maurer, the SVP/UDC and the BDP/PBD hold one seat each.

[59] The Judiciary: The Federal Supreme Court (http://www.ch.ch/behoerden/00215/00329/00353/index.html?lang=en) ch.ch. Retrieved on 2009-12-15

[60] Cantons (http://www.ch.ch/behoerden/00215/00330/index.html?lang=en) ch.ch. Retrieved on 2009-12-15

[61] How direct democracy makes Switzerland a better place (http://www.telegraph.co.uk/news/1435383/How-direct-democracy-makes-Switzerland-a-better-place.html) telegraph.co.uk. Retrieved on 2009-12-04

[62] Since 1999, an initiative can also be in the form of a general proposal to be elaborated by Parliament, but because it is considered less attractive for various reasons, this form of initiative has yet to find any use.

[63] That is a majority of 23 cantonal votes, because the result of the popular vote in the six traditional half-cantons each counts as half the vote of one of the other cantons.

[64] Tremblay; Lecours; et al. (2004) Mapping the Political Landscape. Toronto: Nelson.

[65] Turner; Barry (2001). The Statement's Yearbook. New York: MacMillan Press ltd.

[66] Banks, Arthur (2006). Political Handbook of The World 2005–2006. Washington: Cq Press.

[67] Enclaves of the world (http://enclaves.webs.com/westerneurope.htm) enclaves.webs.com. Retrieved on 2009-12-15

[68] Vorarlberg (http://www.austria.org/content/view/74/) austria.org. Retrieved on 2009-12-15

[69] c2d@zda.uzh.ch (2008-08-20). "unige.ch – Direct democracy in the world" (http://www.c2d.ch/?entit=10&vote=101&lang=). C2d.ch. . Retrieved 2010-04-29.

[70] Neutrality and isolationism (http://www.swissworld.org/en/politics/foreign_policy/neutrality_and_isolationism/) swissworld.org, Retrieved on 2009-06-23

[71] "Switzerland – Country history and economic development" (http://www.nationsencyclopedia.com/economies/Europe/Switzerland-COUNTRY-HISTORY-AND-ECONOMIC-DEVELOPMENT.html). *nationsencyclopedia.com*. . Retrieved 12 December 2009.

[72] Henri Dunant, the Nobel Peace Prize 1901 (http://nobelprize.org/nobel_prizes/peace/laureates/1901/dunant-bio.html) nobelprize.org. Retrieved on 2009-12-02

[73] Sports directory (http://www.ifsports-guide.ch/english/navigation/annuaire_en.html) if-sportsguide.ch. Retrieved on 2010-01-25

[74] An initiative to abandon this practice (http://www.schutz-vor-waffengewalt.ch/) has been launched on 4 September 2007, and supported by GSoA, the Green Party of Switzerland and the Social Democratic Party of Switzerland as well as other organisations which are listed at schutz-vor-waffengewalt.ch (http://www.schutz-vor-waffengewalt.ch/organisationen.html)

[75] "Zwei Drittel der Rekruten diensttauglich (Schweiz, NZZ Online)" (http://www.nzz.ch/nachrichten/schweiz/zwei_drittel_der_rekruten_diensttauglich_1.687233.html). . Retrieved 23 February 2009.

[76] Armeezahlen www.vbs.admin.ch (http://www.vbs.admin.ch/internet/vbs/de/home/documentation/armeezahlen/eff.html) (German)

[77] As context, according to Edwin Reischauer, "To be neutral you must be ready to be highly militarized, like Switzerland or Sweden." – *see* Chapin, Emerson. "Edwin Reischauer, Diplomat and Scholar, Dies at 79," (http://query.nytimes.com/gst/fullpage.html?res=9C0CE2DC163AF931A3575AC0A966958260&scp=1&sq=edwin+o+reischauer&st=nyt) *New York Times*. September 2, 1990.

[78] Volksabstimmung vom 26. November 1989 (http://www.admin.ch/ch/d//pore/va//19891126/index.html) admin.ch. Retrieved on 2010-01-25

[79] *L'évolution de la politique de sécurité de la Suisse* ("Evolution of Swiss Security Policies") by Manfred Rôsch, NATO.int (http://www.nato.int/docu/revue/1993/9306-05.htm)

[80] Volksinitiative 'für eine glaubwürdige Sicherheitspolitik und eine Schweiz ohne Armee (in German) (http://www.admin.ch/ch/d//pore/va/20011202/det482.html) admin.ch. Retrieved on 2009-12-07

[81] Watches (http://www.swissworld.org/en/economy/key_sectors/watches/) swissworld.org. Retrieved on 2009-07-20

[82] 2010 Index of Economic Freedom: Switzerland (http://www.heritage.org/index/country/Switzerland) heritage.org. Retrieved on 2011-01-25

[83] Note: Hong Kong, Guernsey, Cayman Islands and Gibraltar are not sovereign nations. "CIA World Factbook" (https://www.cia.gov/library/publications/the-world-factbook/rankorder/2004rank.html). Cia.gov. . Retrieved 2011-05-29.

[84] "CIA: Switzerland inflation rate" (https://www.cia.gov/library/publications/the-world-factbook/fields/2092.html). Cia.gov. . Retrieved 2011-05-29.

[85] "World Economic Forum – Global Competitiveness Report" (http://www.weforum.org/en/initiatives/gcp/Global Competitiveness Report/index.htm). Weforum.org. . Retrieved 2010-04-29.

[86] Taylor & Francis Group (2002). *Western Europe*. Routledge. pp. 645–646. ISBN 1857431529.

[87] "Zürich – the Highest Quality of Life Worldwide for the Seventh Successive Year" (http://www.zuerich.com/de/Besucher/Information/facts/6-details.html#details). Zuerich.com. . Retrieved 2010-04-29.

[88] "Six Swiss companies make European Top 100" (http://www.swissinfo.ch/eng/business/detail/Six_Swiss_companies_make_European_Top_100.html?siteSect=161&sid=7174196&cKey=1161172317000). swissinfo.ch. 18 October 2008. . Retrieved 22 July 2008.

[89] Swiss Statistical Yearbook 2008 by Swiss Federal Statistical Office

[90] Swiss jobless reach 12-year high – a mere 4.4 pct (http://www.businessweek.com/ap/financialnews/D9D3FGVO0.htm)

[91] Policy Brief: Economic Survey of Switzerland, 2007 (http://www.oecd.org/dataoecd/39/8/39539300.pdf) (326 KiB), OECD

[92] Economic Policy Reforms: Going for Growth 2008 – Switzerland Country Note (http://www.oecd.org/dataoecd/29/49/40202407.pdf) (45 KiB)

[93] Domestic purchasing power of wages (http://www.osec.ch/internet/osec/en/home/invest/factors/infrastructure/live/costs.-RelatedBoxSlot-47301-ItemList-89920-File.File.pdf/Domestic purchasing power of wages E.pdf) (68 KiB)

[94] Switzerland tops in buying power (http://goliath.ecnext.com/coms2/gi_0199-4179358/Switzerland-tops-in-buying-power.html) goliath.ecnext.com. Retrieved on 2010-01-14.

[95] Want the world's best wages? Move to Switzerland (http://www.reuters.com/article/idUSTRE57Q23H20090827) reuters.com. Retrieved on 2010-01-14.

[96] The Swiss education system (http://www.swissworld.org/en/education/general_overview/the_swiss_education_system/) swissworld.org, Retrieved on 2009-06-23

[97] ETH Zurich (Swiss Federal Institute of Technology) (http://www.topuniversities.com/university/199/eth-zurich-swiss-federal-institute-of-technology) topouniversities.com. Retrieved on 2010-04-30

[98] Top.Universities (http://www.topuniversities.com/world-university-rankings) Retrieved on 2010-04-30

[99] In 2008, the ETH Zurich was ranked 15th in the field *Natural Sciences and Mathematics* by the Shanghai Academic Ranking of World Universities and the EPFL in Lausanne was ranked 18th in the field *Engineering/Technology and Computer Sciences* by the same ranking.

[100] "Shanghai Ranking 2008 Top 100 world universities in Natural Sciences and Mathematics" (http://ed.sjtu.edu.cn/ARWU-FIELD2008/SCI2008.htm). Ed.sjtu.edu.cn. . Retrieved 2010-11-02.

[101] Education at Glance 2005 (http://www.ecs.org/html/offsite.asp?document=http://www.oecd.org/dataoecd/20/25/35345692.pdf) by the OECD: Percentage of foreign students in tertiary education.

[102] Dr. phil. Roland Mueller. "Muellerscience.com" (http://www.muellerscience.com/SPEZIALITAETEN/Schweiz/SchweizerNobelpreistraeger.htm). Muellerscience.com. . Retrieved 2011-05-29.

[103] Nobel prizes in non-science categories included.

[104] "Mueller Science – Spezialitaeten: Schweizer Nobelpreisträger" (http://www.muellerscience.com/SPEZIALITAETEN/Schweiz/SchweizerNobelpreistraeger.htm). . Retrieved 31 July 2008.

[105] info.cern.ch (http://info.cern.ch/) Retrieved on 2010-04-30

[106] "CERN - the largest laboratory in the world www.swissworld.org" (http://www.swissworld.org/en/switzerland/resources/story_switzerland/cern_the_largest_laboratory_in_the_world/). Swissworld.org. . Retrieved 2010-04-29.

[107] "How to get to CERN" (http://user.web.cern.ch/user/Institute/Integration/GetToCERN/GetToCERN.html). .

[108] Company overview www.oerlikon.com (http://www.oerlikon.com/ecomaXL/index.php?site=SPACE_EN_company_overview)

[109] maxon motor ag (2004-01-04). "Media releases" (http://www.maxonmotor.ch/ch/en/media_releases_5619.html). Maxonmotor.ch. . Retrieved 2010-04-29.

[110] Prof Clive Church (may 2003). "The contexts of Swiss opposition to Europe" (http://www.webcitation.org/5lZameh86) (PDF, 124 KiB). Sussex European Institute. p. 12. Archived from the original (http://www.sussex.ac.uk/sei/documents/wp64.pdf) on 2009-11-26. . Retrieved 13 June 2008.

[111] "Volksinitiative "Ja zu Europa!" *(Initiative "Yes to Europe!")"* (http://www.bfs.admin.ch/bfs/portal/de/index/themen/17/03/blank/key/2001/01.Document.22675.pdf) (in German) (PDF, 1.1 MiB). BFS/OFS/UST. 13 February 2003. . Retrieved 15 June 2008.

[112] "Volksinitiative "Ja zu Europa!", nach Kantonen. (Initiative "Yes to Europe!" by Canton)." (http://www.bfs.admin.ch/bfs/portal/de/index/themen/17/03/blank/key/2001/01.Document.85488.xls) (in German) (XLS). BFS/OFS/UST. 16 January 2003. . Retrieved 15 June 2008.

[113] Switzerland and the European Union (http://www.europa.admin.ch/themen/00499/00503/index.html?lang=en) europa.admin.ch. Retrieved on 2010-01-25

[114] Switzerland in Schengen: end to passport checks (http://www.euronews.net/2008/12/11/switzerland-in-schengen-end-to-passport-checks/) euronews.net. Retrieved on 2010-01-25

[115] "Vote No. 502 – Summary" (http://www.admin.ch/ch/d/pore/va/20030518/det502.html) (in German). 18 May 2003. .

[116] "Vote No. 501 – Summary" (http://www.admin.ch/ch/d/pore/va/20030518/det501.html) (in German). 18 May 2003. .

[117] "Federal government energy research" (http://www.bfe.admin.ch/forschungnetze/01223/index.html?lang=en). 16 January 2008. .

[118] "Switzerland to Phase Out Nuclear Energy; E.U. Strikes Deal on 'Stress Tests'" (http://news.sciencemag.org/scienceinsider/2011/05/switzerland-to-phase-out-nuclear.html). 25 May 2011. .

[119] "Verkehrsleistungen – Daten, Indikatoren admin.ch (German)" (http://www.bfs.admin.ch/bfs/portal/de/index/themen/11/05/blank/key/verkehrsleistungen/mengen.html). Bfs.admin.ch. 2010-03-25. . Retrieved 2010-04-29.

[120] Schienenverkehr (http://www.bav.admin.ch/dokumentation/publikationen/00475/01623/01624/index.html?lang=de) admin.ch (German)

[121] Rhaetian Railway in the Albula / Bernina Landscapes (http://whc.unesco.org/en/list/1276/) unesco.org

[122] Swiss sit atop ranking of greenest nations (http://www.msnbc.msn.com/id/22802081/) msnbc.com. Retrieved on 2009-12-02

[123] Party grouping (http://unfccc.int/parties_and_observers/parties/negotiating_groups/items/2714.php) unfccc.int. Retrieved on 2009-12-02

[124] W3design. "Swiss Recycling" (http://www.swissrecycling.ch/deutsch/wregel.htm). Swissrecycling.ch. . Retrieved 2010-04-29.

[125] Stadtreinigung Basel-Stadt (http://www.stadtreinigung-bs.ch/page.php?lang=0&sel=114)—Pricelist bags and stickers

[126] "Recycling around the world" (http://news.bbc.co.uk/1/hi/world/europe/4620041.stm). BBC. 25 June 2005. . Retrieved 24 April 2006.

[127] Richtig Entsorgen (Kanton Basel-Stadt) (http://www.stadtreinigung-bs.ch/data/0d1b64Sauberbuch2004.pdf) (1.6 MiB)—Wilde Deponien sind verboten... Für die Beseitigung widerrechtlich deponierter Abfälle wird zudem eine Umtriebsgebühr von Fr. 200.– oder eine Busse erhoben (page 90)

[128] History of paper manufacturing (http://www.geschichteinchronologie.ch/welt/papiergeschichte.html) in German, retrieved 03-05-2011

[129] Swiss Federal Statistical Office. "Languages and religions – Data, indicators" (http://www.bfs.admin.ch/bfs/portal/en/index/themen/01/05/blank/key/sprachen.html). . Retrieved 9 October 2007. The first number refers to the share of languages within total population. The second refers to the Swiss citizens only.

[130] The Parliamentary Services (http://www.parlament.ch/e/service-presse/parlamentsdienste/aufgaben/Pages/default.aspx) parlament.ch. Retrieved on 2010-01-25

[131] "SBB: Billette – OnlineTicket" (http://mct.sbb.ch/mct/reisemarkt/billette/online-ticket.htm). Mct.sbb.ch. . Retrieved 2010-04-29.

[132] Minorities and bilingualism (http://www.swissworld.org/en/people/language/minorities_and_bilingualism/) swissworld.org. Retrieved on 2010-01-25

[133] Ausländerinnen und Ausländer in der Schweiz – Bericht 2008 (German) (http://www.bfs.admin.ch/bfs/portal/de/index/themen/01/22/publ.Document.114724.pdf) (1196 KiB), Swiss Federal Statistical Office, page 12.

[134] Ausländerinnen und Ausländer in der Schweiz – Bericht 2008 (German) (http://www.bfs.admin.ch/bfs/portal/de/index/themen/01/22/publ.Document.114724.pdf) (1196 KiB), Swiss Federal Statistical Office, page 72.

[135] Foreign population in Switzerland detailed by nationality, 1980–2006 (German) (http://www.bfs.admin.ch/bfs/portal/de/index/themen/01/07/blank/key/01/01.Document.67321.xls), Swiss Federal Statistical Office.

[136] Definitive report on racism in Switzerland by UN expert (http://www.humanrights.ch/home/en/Switzerland/Policy/Racism/Studies/idart_5119-content.html) humanrights.ch

[137] Switzerland (http://www.who.int/countries/che/en/index.html) who.int. Retrieved on 2009-06-29

[138] Life expectancy at birth, 2006 (http://apps.who.int/whosis/database/country/compare.cfm?strISO3_select=CHE&strIndicator_select=LEX0Male,LEX0Female&language=english&order_by=FirstValue DESC) who.int. Retrieved on 2009-06-29

[139] OECD Health Data 2006 (http://www.oecd.org/dataoecd/29/52/36960035.pdf) oecd.org. Retrieved on 2009-06-29

[140] OECD and WHO survey of Switzerland's health system (http://www.oecd.org/document/47/0,2340,en_2649_201185_37562223_1_1_1_1,00.html) oecd.org. Retrieved on 2009-06-29

[141] Where people live (http://www.swissworld.org/en/geography/town_and_country_planning/where_people_live/) swissworld.org. Retrieved on 2009-06-26

[142] Städte und Agglomerationen unter der Lupe (http://www.are.admin.ch/dokumentation/00121/00224/index.html?lang=de&msg-id=27412) admin.ch. Retrieved on 2009-06-26

[143] Swiss countryside succumbs to urban sprawl (http://www.swissinfo.ch/eng/front/Swiss_countryside_succumbs_to_urban_sprawl.html?siteSect=106&sid=9823369&cKey=1223485367000&ty=st) swissinfo.ch. Retrieved on 2009-06-30

[144] Enquête représentative sur l'urbanisation de la Suisse (Pronatura) (http://www.gfs-zh.ch/content.php?pid=201) gfs-zh.ch. Retrieved on 2009-06-30

[145] Swiss plateau (http://www.swissworld.org/en/geography/the_three_regions/the_swiss_plateau/) swissworld.org. Retrieved on 2009-06-29

[146] Quality of living (http://www.mercer.com/qualityofliving) mercer.com. Retrieved on 2009-06-26

[147] International Religious Freedom Report 2004 – Switzerland (http://www.state.gov/g/drl/rls/irf/2004/35487.htm), U.S. Department of State.

[148] "CIA World Factbook section on Switzerland" (https://www.cia.gov/library/publications/the-world-factbook/geos/sz.html#People). Cia.gov. . Retrieved 2011-05-29.

[149] Bovay, Claude; Raphaël Broquet (December 2004). "Introduction" (http://www.bfs.admin.ch/bfs/portal/fr/index/news/publikationen.Document.50517.pdf) (in French) (PDF). *Recensement fédéral de la population 2000*. Neuchâtel: Federal Statistical Office. p. 132. ISBN 3-303-16074-0. . Retrieved 21 August 2010

[150] Social values, Science and Technology (http://ec.europa.eu/public_opinion/archives/ebs/ebs_225_report_en.pdf)PDF (1.64 MiB), Eurobarometer, June 2005.

[151] Greeley, Andrew. 2003. Religion in Europe at the End of the Second Millennium. New Brunswick, New Jersey: Transaction Publishers

[152] Reclus, Élisée (1881). *The Earth and Its Inhabitants*. D. Appleton and Company. p. 478.
[153] Volksabstimmung vom 2. März 1980 (http://www.admin.ch/ch/d/pore/va/19800302/index.html) admin.ch. Retrieved on 2010
[154] Swiss culture (http://www.swissworld.org/en/culture/) swissworld.org. Retrieved on 2009-12-01
[155] European Year of Intercultural Dialogue (http://delche.ec.europa.eu/fileadmin/user_upload/Homepage_Dokumente/Presse_und_Information/Reden/2008-11-13_European_Year_of_Intercultural_Dialogue_2008__AFS_Intercultural_Programmes_Kaefigturm.pdf) Dr Michael Reiterer. Retrieved on 2009-12-01
[156] Switzerland: culture (http://www.traveldocs.com/ch/culture.htm) traveldocs.com. Retrieved on 2009-12-01
[157] Museums (http://www.swissworld.org/en/leisure/relaxation/museums/) swissworld.org. Retrieved on 2009-12-02
[158] Lucerne Festival (http://topics.nytimes.com/top/reference/timestopics/subjects/l/lucerne_festival/index.html) nytimes.com. Retrieved on 2010-12-15
[159] Switzerland's Prestigious Music Festival-Montreux Jazz Festival (http://www.nowpublic.com/culture/switzerland-s-prestigious-music-festival-montreux-jazz-festival) nowpublic.com. Retrieved on 2009-12-02
[160] Film festivals (http://www.swissworld.org/en/culture/film/film_festivals/) swissworld.org. Retrieved on 2009-12-02
[161] Mountains and hedgehogs (http://www.swissworld.org/en/culture/swissness/mountains_and_hedgehogs/) swissworld.org. Retrieved on 2009-12-01
[162] In Search of Natural Identity: Alpine Landscape and the Reconstruction of the Swiss Nation (http://journals.cambridge.org/download.php?file=/CSS/CSS40_04/S0010417598001686a.pdf&code=f5973a22cda752eed69390f92ddf4fcd), Oliver Zimmer, London School of Economics and Political Science
[163] Folk music (http://www.swissworld.org/en/culture/music/folk_music/) swissworld.org. Retrieved on 2009-12-02
[164] Culture of Switzerland (http://www.europe-cities.com/en/648/switzerland/culture/) europe-cities.com. Retrieved on 2009-12-14
[165] Art in literature (http://www.cp-pc.ca/english/switzerland/) cp-pc.ca. Retrieved on 2009-12-14
[166] From Encyclopædia Britannica Eleventh Edition, Swiss literature
[167] Literature (http://www.swissworld.org/en/culture/literature/german_speaking_authors/) swissworld.org, Retrieved on 2009-06-23
[168] Press and the media (http://www.ch.ch/private/00085/00090/00479/00480/index.html?lang=en) ch.ch. Retrieved on 2009-06-25
[169] Press in Switzerland (http://www.pressreference.com/Sw-Ur/Switzerland.html) pressreference.com. Retrieved on 2009-06-25
[170] Sport in Switzerland (http://www.europe-cities.com/en/672/switzerland/sport/) europe-cities.com. Retrieved on 2009-12-14
[171] A brief history of bobsleigh (http://www.fibt.com/index.php?id=39) fibt.com. Retrieved on 2009-11-02
[172] "IIHF World Championships 2009 official website" (http://www.iihf.com/channels/iihf-world-championship-oc09/home-oc/tournament-information.html). Iihf.com. 2009-05-10. . Retrieved 2010-04-29.
[173] http://www.hockeyarenas.net/index.php3?ctID=ch&size=0&page=0701&c=eu&ctID=eu&atType=0&show=25&tdSaison=2011
[174] Roger Federer's Grand Slam Titles (http://sportsillustrated.cnn.com/vault/gallery/featured/GAL1157546/1/index.htm) sportsillustrated.cnn.com. Retrieved on 2010-06-14
[175] "Switzerland lifts ban on motor racing" (http://en.wikinews.org/wiki/Switzerland_lifts_ban_on_motor_racing). GrandPrix.com & DueMotori.com. 6 June 2007. . Retrieved 23 September 2008.
[176] Hornussen (http://www.swissroots.org/swissroots/en/stories/heritage/Heritage/Swiss Customs/Hornussen Baseball Swissstyle) swissroots.org. Retrieved on 2010-01-25
[177] Tradition and history (http://www.interlaken.ch/erlebnisse/sommer/festivals/unspunnenfest/tradition-und-geschichte.html?L=3) interlaken.ch. Retrieved on 2010-01-25
[178] Flavors of Switzerland (http://www.theworldwidegourmet.com/countries/flavors-of-switzerland/) theworldwidegourmet.com. Retrieved on 2009-06-24
[179] Michelin Guide Switzerland 2010 attests to the high quality of gourmet cooking with one new 2 star restaurant and 8 new one star (http://servicesv2.webmichelin.com/frontnews/servlet/GetElement?elementCode=56670) Press information, Michelin. Retrieved on 2009-12-14
[180] Swiss region serves up food with star power (http://www.usatoday.com/travel/destinations/2005-04-26-swiss-food_x.htm) usatoday.com. Retrieved on 2009-12-14
[181] Chocolate (http://www.swissworld.org/en/switzerland/swiss_specials/swiss_chocolate/swiss_breakthroughs/) swissworld.org. Retrieved on 2009-06-24
[182] Swiss Chocolate (http://www.germanworldonline.com/index.php/swisschocolatehistory/) germanworldonline.com. Retrieved on 2010-06-14
[183] Wine-producing Switzerland in short (http://www.swisswine.ch/english/bienv/main.asp) swisswine.ch. Retrieved on 2009-06-24
[184] Table 38. Top wine consuming nations per capita, 2006 (http://www.winebiz.com.au/statistics/world.asp) winebiz.com. Retrieved on 2010-06-14

References

Bibliography

- Church, Clive H. (2004) *The Politics and Government of Switzerland*. Palgrave Macmillan. ISBN 0-333-69277-2.
- Dalton, O.M. (1927) *The History of the Franks, by Gregory of Tours*. Oxford: The Clarendon Press.

- Fahrni, Dieter. (2003) *An Outline History of Switzerland. From the Origins to the Present Day.* 8th enlarged edition. Pro Helvetia, Zürich. ISBN 3-908102-61-8
- Historical Dictionary of Switzerland (2002–). Published electronically and in print simultaneously in three national languages of Switzerland.

External links

Government

- The Federal Authorities of the Swiss Confederation (http://www.admin.ch/)
- Chief of State and Cabinet Members (https://www.cia.gov/library/publications/world-leaders-1/world-leaders-s/switzerland.html)
- Swissworld.org, Switzerland's information portal (http://www.swissworld.org/)
- Swiss Statistics (http://www.bfs.admin.ch/bfs/portal/en/index.html), official website of the Swiss Federal Statistical Office

Reference

- Switzerland (https://www.cia.gov/library/publications/the-world-factbook/geos/sz.html) entry at *The World Factbook*
- Switzerland (http://www.britannica.com/EBchecked/topic/577225/Switzerland) entry at *Encyclopædia Britannica*
- Switzerland (http://ucblibraries.colorado.edu/govpubs/for/switzerland.htm) at *UCB Libraries GovPubs*
- Switzerland (http://www.dmoz.org/Regional/Europe/Switzerland/) at the Open Directory Project

Geography

- Wikimedia Atlas of Switzerland
- Switzerland travel guide from Wikitravel
- Federal Office of Topography (http://www.swisstopo.ch/)
- Searchable interactive map (search.ch) (http://map.search.ch/)

Travel

- MySwitzerland.com (http://www.myswitzerland.com/), official website of Switzerland Tourism
- Worldofmaps.net (http://www.worldofmaps.net/en/europe/map-switzerland.htm), Online Maps of Switzerland

History

- Historical Dictionary of Switzerland (http://www.hls-dhs-dss.ch/index.php) **(German)** **(French)** **(Italian)**
- History of Switzerland (http://history-switzerland.geschichte-schweiz.ch/index.html)
- Swiss American Historical Society (http://swissamericanhistoricalsoc.org/)

Languages

- swiss-linguistics.com (http://www.swiss-linguistics.com/) Information portal on current linguistic research in Switzerland

News media

- Neue Zürcher Zeitung (http://nzz.ch/) **(German)**, a Swiss daily newspaper
- Le Temps (http://www.letemps.ch/) **(French)**, a Swiss daily newspaper
- Corriere Del Ticino (http://www.cdt.ch/) **(Italian)**, a Swiss daily newspaper
- swissinfo.ch, Swiss News – Worldwide (http://www.swissinfo.ch/)
- expatinch.com, Swiss media resources page (http://expatinch.com/html/media_tv_telephony.html)

Education

- Swiss Universities Handbook – Education in Switzerland (http://www.universitieshandbook.com/)

- Universities in Switzerland (http://www.swissuniversity.ch/)
- The Swiss School System (http://www.educa.ch/)
- sieps.ch (http://www.sieps.ch/) Information Services on Swiss Private Schools and Universities

Science, research, and technology

- The Swiss Portal for Research and Innovation (http://www.myscience.ch/)
- State Secretariat for Education and Research, SER (http://www.sbf.admin.ch/)
- The Swiss National Science Foundation (http://wwww.snf.ch/)

kbd:Шуэцарэ gag:Şveyţariya koi:Швейцарму ltg:Šveicareja pfl:Schwaiz rue:Швайцарія xmf:შვეიცარია

Aar

Aar

The Aar at Bern

Drainage basin of the Aar

Origin	Canton of Bern, Switzerland
Mouth	Rhine
Basin countries	Switzerland, France
Length	295 kilometres (183 mi)
Basin area	17779 km^2 (6865 sq mi)

The **Aar** (German *Aare*), a tributary of the High Rhine, is the longest river that both rises and ends entirely within Switzerland.[1]

Its total length from its source to its junction with the Rhine comprises about 295[2] km (183 miles),[1] during which distance it descends 1565 m (5135 ft), draining an area of 17779 km^2 (6865 sq mi), including the whole of central Switzerland.[2]

Course

The Lower Aar Glacier

The Aar rises in the great Aar Glaciers of the Bernese Alps, in the canton of Bern and west of the Grimsel Pass.[1] It runs east to the Grimsel Hospice, below the Finsteraarhorn, and then northwest through the Haslital, forming on the way the magnificent Handegg Waterfall, 46 m (151 ft), past Guttannen, and piercing the limestone barrier of the Kirchet by a major canyon, before reaching Meiringen, situated on a plain. A little past Meiringen, near Brienz, the river expands into Lake Brienz where it becomes navigable. Near the west end of the lake it receives its first important tributary, the Lütschine. It then runs across the swampy plain of the Bödeli between Interlaken and Unterseen before flowing into Lake Thun.[1]

Near the west end of Lake Thun, the river receives the waters of the Kander, which has just been joined by the Simme. On flowing out of the lake it passes through Thun, and then flows through the city of Bern, passing beneath eighteen bridges and around the steeply-flanked peninsula on which the Old City is located. The river soon changes its northwesterly flow for a due westerly direction, but after receiving the Saane or Sarine it turns north until it nears Aarberg. There, in one of the major Swiss engineering feats of the 19th century, the Jura water correction, the river, which had previously rendered the countryside north of Bern a swampland through frequent flooding, was diverted by the Hagneck Canal into Lake of Bienne. From the upper end of the lake the river issues through the Nidau-Büren channel and then runs east to Büren. The lake absorbs huge amounts of eroded gravel and snowmelt that the river brings from the Alps, and the former swamps have become fruitful plains: they are known as the "vegetable garden of Switzerland".

The Aar at Innertkirchen

From here the Aar flows northeast for a long distance, past the ambassador town Solothurn[1] (below which the Grosse Emme flows in on the right), Aarburg (where it is joined by the Wigger), Olten, Aarau,[1] near which is the junction with the Suhre, and Wildegg, where the Hallwiler Aa falls in on the right. A short distance further, below Brugg it receives first the Reuss, and shortly afterwards the Limmat. It now turns due north, and soon becomes itself an tributary of the Rhine, which it surpasses in volume when the two rivers unite at Koblenz (Switzerland), opposite Waldshut, Germany. The Rhine, in turn, empties into the North Sea after crossing into the Netherlands.

Aar Gorge

The Aar Gorge (*Aareschlucht*) is a section of the river that carves through a limestone ridge near the town of Meiringen.

Inside the Aar Gorge

Tributaries

- Limmat
  - Sihl
  - Linth (main tributary of Lake Zürich)
- Reuss (*see also: Reuss (river)#Tributaries*)
  - Kleine Emme
  - Sarner Aa
  - Engelberger Aa
  - Schächen
  - Furkareuss
- Suhre
- Wigger
- Emme
- Zihlkanal
  - Suze
  - Broye
  - Orbe
- Saane (Sarine)
  - Sense
- Kander
  - Simme
  - Allenbach
- Lütschine

At the "Wasserschloss", where the rivers Aar, Reuss and Limmat come together

The convergence of the Aar and the Rhine at Koblenz

Reservoirs

- Lake Wohlen, 481 m
- Niederriedsee, 461 m
- Klingnauer Stausee, 318 m

History

The river's name may be attested to in the Gaulish (Helvetic) Berne zinc tablet which dates back to Roman Gaul. Its Roman name was *Obringa*.

References

[1] Bridgwater, W. & Beatrice Aldrich. (1966) *The Columbia-Viking Desk Encyclopedia.* Columbia University. p. 11.
[2] "Aare River." (http://search.eb.com/eb/article-9003206) *Encyclopædia Britannica.* 9 January 2007

External links

- The Aar Gorge (Aareschlucht) (http://www.aareschlucht.ch)

Article Sources and Contributors

Klingnauer Stausee *Source*: http://en.wikipedia.org/w/index.php?title=Klingnauer_Stausee *Contributors*: D6, Dbachmann, Docu, Grutness, LrdChaos, Manoillon, MichaelFrey, Numbo3, 5 anonymous edits

Böttstein *Source*: http://en.wikipedia.org/w/index.php?title=B%C3%B6ttstein *Contributors*: BillC, Brewcrewer, D6, DidiWeidmann, Docu, Dr. Blofeld, Jwnabd, Ksnow, MichaelFrey, Schweizerfreund, Stemonitis, Tobyc75, Wetman, Woohookitty, 5 anonymous edits

Zurzach District *Source*: http://en.wikipedia.org/w/index.php?title=Zurzach_District *Contributors*: BillC, Bobblewik, Bobet, Brewcrewer, D6, DidiWeidmann, Jeepday, Karl-Henner, Markko, Od Mishehu, Pumpie, Reza1615, Saltose, Schwyz, Tobyc75, Voyager, 8 anonymous edits

Baden, Switzerland *Source*: http://en.wikipedia.org/w/index.php?title=Baden%2C_Switzerland *Contributors*: Absolwent, Adlei86, Andre Engels, Badener, Baldhur, BillC, Bob Burkhardt, Chl, D6, DF08, Dbachmann, Docu, Doulos Christos, Dribblingscribe, Gamonetus, GiovanniS, Grahamec, Harry Wood, Ikiwaner, Irmgard, Jevansen, Jezerfetnae, Jwnabd, Kauczuk, Kjkolb, Kokiri, Ksnow, Ktsquare, Lupin, Lylum, MadGeographer, Marek69, Mic, Michael Hardy, Mikaey, Mirv, Monkbel, Numbo3, Olivier, Qolume, Rich Farmbrough, Roland zh, Sandstein, Seanmullan, Sluzzelin, Snowbunni, ThomasAndrewNimmo, Tobyc75, Ulric1313, Viperch, Voyager, Wilchett, 19 anonymous edits

La Tène culture *Source*: http://en.wikipedia.org/w/index.php?title=La_T%C3%A8ne_culture *Contributors*: 83d40m, AN(Ger), AdRock, Amkilpatrick, Andycjp, Angr, Athinaios, Baronnet, Bastin, Biruitorul, Blackcats, Bobblewik, Bogdangiusca, Caspian blue, Claude girardin, Codrinb, Conversion script, Creidieki, CristianChirita, Cuchullain, Cunibertus, Cépey, Dbachmann, Deacon of Pndapetzim, Dougweller, Eleassar, Fortdj33, Freakofnurture, Fuzzypeg, Gene Nygaard, Glaurung, Glst2, Hairy Dude, Hede2000, Hmains, Hurlzzz, Ipigott, Ischa2, Jamelan, Jguk 2, Johnbod, Joy, Kazakhstan rocks, Kiyote23, Marco polo, Mhocker, Michael Hardy, Mitrius, Mzajac, NYArtsnWords, Neddyseagoon, Nicknack009, Olivier, OlofE, Oxymoron83, POUMPOUM, Per Honor et Gloria, PerlKnitter, Pigman, Prof saxx, Pymouss, Qatter, QuartierLatin1968, RHole, RIH-V, RedWolf, Reiner Martin, Rich Farmbrough, Rjm at sleepers, Rodw, Sam Hocevar, Sandstein, Sdrivera42, Slovenski Volk, Sluzzelin, Sparspost, Stbalbach, StefanoC, Suitcivil133, Summer Song, Sympleko, Tabletop, Tarnas, Tarquin, Tartessos75, The Minister of War, Thebrainkid, Theo10011, Tobyc75, Toyokuni3, Trainthh, Trigaranus, Vald, VerkaS, Wachowich, Wclark, Wetman, Woohookitty, Yak, 49 anonymous edits

Aargau *Source*: http://en.wikipedia.org/w/index.php?title=Aargau *Contributors*: -- April, 2Neuenburg, Adlei86, Ahoerstemeier, AlexAnnecy, Alfirin, Andre Engels, Aphaia, Aram33, BajanZindy, Baldhur, Bastin, Bob Burkhardt, Bobblewik, Boleyn, Claude girardin, Conversion script, D6, David Newton, Dbachmann, Dcoetzee, Den fjättrade ankan, Denelson83, Docu, Dr. Blofeld, Dribblingscribe, DwightKingsbury, Editore99, Fastifex, Funandtrvl, Funnyhat, Gilded Lily, Goldfishbutt, Gwern, Habib87, Harleyrules170, Howcheng, Ike, Innotata, JM.Beaubourg, Japanese Searobin, Jauhienij, Jbyrde1, Jezerfetnae, KingTT, KirkEN, Klapper, Kokiri, Kummi, Lawrencekhoo, LeoNomis, Lightmouse, Lincher, Longbow4u, Macronyx, Magnum1, Marek69, Matt Gies, Maximaximax, Mgimpel, Mirv, Mowsbury, Nancy, Neuenburg, Numbo3, Olivier, OwenBlacker, Ratcliffe mike, Reisio, Rich Farmbrough, Robert Horning, Rocastelo, SWAdair, Saltose, Scipius, Seanmullan, Shabbs, Sidious1701, Silverhelm, Sluzzelin, Smilo Don, Snori, Snowdog, Synthetik, Tobyc75, Trilobite, Vanka5, Varoon Arya, Voyager, Walden69, Wasbeer, WereSpielChequers, Wetman, Wik, Wikibob, Will Beback, Wknight94, Ze miguel, 69 anonymous edits

Switzerland *Source*: http://en.wikipedia.org/w/index.php?title=Switzerland *Contributors*: (jarbarf), -- April, 0, 05, 100110100, 130.60.153.xxx, 159753, 16@r, 334a, 5dsddddd, 661kts, 7anonymous7, 7tests, 80sfaan, A Macedonian, A-giau, A3RO, AB-me, ABF, AFA, Aaalroex, Aaker, Aaron Einstein, Aaronbrick, Abc518, Abrech, AbsoluteZero280, Academic38, Acalamari, Access Denied, Acelor, Achtzehn, Acs4b, Adam Bishop, Adam78, Adamahill, Adashiel, Adhall, Adi4094, Aetheling, Aff123a, Agapius1987, Aguasde13013, Ahoerstemeier, Ahuskay, Airplane yow, Aitias, AkAiBaRa, Akanemoto, Alan16, Alansohn, Aleph4, Alethiareg, Alex S, Alex.muller, Alex45, Alexander Domanda, AlexiusHoratius, Ali K, Alivath, Alperen, Alphabravotango, Alphasinus, AltWiki, Altermike, Ambg, American1991, Amerika, Amorim Parga, Andajew, Andre Engels, Andrew Levine, Andrew33base, Andrewlp1991, Andrewrp, Android Mouse, Andy M. Wang, Andy Marchbanks, Andyindia, AngelIsaac, AngelOfSadness, Anger22, Angr, Angusmclellan, Animum, Ann Stouter, Anna Lincoln, AnnaFrance, Anonymous Dissident, Anonymous anonymous, Answar, Antandrus, Antonio Gonzales, AntonioMartin, Anxa, Aoibheannniamh, Apiarius Flavius, Arad, Aranherunar, Armando, Arminescher, Army1987, Arne List, Arwel Parry, Ascend, Atitarev, Atumbabod, AuntFlo, Austin512, Austinthebookworm, Avala, Avenged Eightfold, AvicAWB, Avicennasis, AxG, Axeman89, Ayrton Prost, Azazello, B. Phillips, BAAKO, BBT2005, BRG, BTLizard, Bababoef, Backslash Forwardslash, Badenoch, Bakerccm, BalkanFever, Bambuway, Bananaclaw, Banpei, Barek, Basawala, Baseball Watcher, Baseballbaker23, Bazonka, Bcnviajero, Bdbfly1, Beano, Behemoth, Beland, Bellahdoll, Bender235, Benjiwolf, Benkok101, Bennylin, Benyon3, Bevo74, Bggoldie, Bhumiya, Bickeringwife, Bidabadi, Bigfujitsu, Bilko yes, BillC, Binary TSO, Bjarki S, Blanchardb, Blavertu, BlazeJag, Bleach3, Blessthishouse, Bloody resident, Blue520, BlueMars, BlueNovember, BlueXen, Blueboy96, Bluee Mountain, Bluerasberry, Bmoq, BoLingua, Bob247, BobIV, Bobblewik, Bobbo, Bobg123, Bobo192, Bodnotbod, Bogey97, Bokpasa, Bolo1729, Bongwarrior, Booksworm, Boost3000, BorgQueen, Borisblue, Bosonic dressing, Bourquie, Bradoni, Brainboy109, Brandon, Branlon, BrendelSignature, BrianGV, BrigadierGeneralBaptisms, Brighterorange, Brion VIBBER, Brisvegas, BritishWatcher, Broha8, Brrryan, Brunton, Brutaldeluxe, Bsadowski1, Buaidh, Bucephalus, Bud467, Bugnon, BuickCenturyDriver, Bunchofgrapes, Bunny-chan, C mon, C0nanPayne, C28martin, CIreland, CO, CONNOR-WALSH-95, CRE is king, CSZero, Cae prince, Caffein, CalJW, Calabraxthis, Calbaer, Callum389, Calmargulis, Caltas, Cam147, CambridgeBayWeather, Camilo Sanchez, Can't sleep, clown will eat me, CanadianLinuxUser, Canderra, Canley, Cantanameras, Canterbury Tail, Cantus, CapitalR, CapnPrep, Caponer, Capricorn42, Captain Spleen, CardinalDan, CaseyPenk, Casper2k3, Cate, Catfish Jim and the soapdish, Catfishy456, Caumasee, Causa sui, Cbowers@kumc.edu, CdaMVvWgS, Cdc, CeeWhy2, Cenarium, Centrx, Cepal, Chanakal, Charvex, Chatfecter, Chiacomo, Chochopk, Chris Rocen, Chris the speller, Chrisch, Chrisieboy, Christianbrewerrox, Christopher Sundita, Chrono1084, Chronodm, CiTrusD, CieloEstrellado, Cinephile, City of Destruction, Cjrs 79, ClamIAm, Claude girardin, Cleduc, Clorox, Closedmouth, Clpda, Coffee, Cogito ergo sumo, Cokepepsi, Colonies Chris, Colt .55, Coltfan009, Commander Keane, Commando502, Common appeal, CommonsDelinker, Conglacius, Connormah, Conorbrady.ie, Contributor777, Conversion script, Coolgirl1934, Coolhawks88, Coppernut, Cordless Larry, Corti, Corvus13, Costesseyboy, Cptnono, Crazyjesenia, Crazywaters, Credema, Cremepuff222, Creol, Crhottie01, Cristian Cappiello, Cruzian, Cryptic, Cs-wolves, Cs219cheater, Csabi911, Css, Cub68134, Cunado19, Cuthbertgrant, CyberPanda, Cybercobra, Cyde, Cyhawk, CyrilleDunant, D, D4RK-L3G10N, D6, DG, DJ Clayworth, DJSoundBored, DMacks, DW, DaDoc540, Daa89563, Dacrazysmoose, Dahveed323, Dale-swiss, Dalstadt, Dancter, Danger, Danno uk, Danny, Dapete, Darrendeng, Darthdj31, Davewild, David Gerard, David Kernow, David Liuzzo, David.Monniaux, DavidHolden, DavidNw, Davidfreund, Dawright12, Dayubcpd, Dbachmann, Dbenbenn, Dbfirs, De728631, DeadEyeArrow, Deadstar, Deathphoenix, Dedee73, Deh, Dekisugi, Delpino, Deltabeignet, Deltasigma, Demmy, Denelson83, Deor, Der Eberswalder, DerHexer, Dexter prog, DiENass, DiamondShade, DieZurcher, Diego pmc, Diftong, Dimitrii, Diothoir, Directorstratton, Discospinster, DivineAlpha, Dmn, Doc Quintana, Doc glasgow, DocWatson42, Doctorevil64, Docu, Dodo von den Bergen, Dogfish Jim and the Dixoap, Don Durandal, Donkeydick11, Dookymonkey, Dorftrottel, Dougofborg, Dr. Blofeld, DrYak, Draconion devil, Dragonlord kfb, Drbug, Dreadstar, Dream Focus, Drepanopulos, Drew R. Smith, DriveMySol, Drmies, DrunkenSmurf, Dschwen, Dspradau, Dunkedun, Dust Filter, Dwo, Dwunderli, Dysepsion, E Pluribus Anthony, E Pluribus Anthony redux, E2eamon, ECube Global, EJF, ERcheck, ESkog, EZ, Eaefremov, EamonnPKeane, Eclecticology, EdBever, Edificia, EeepEeep, Efghij, Eggman64, Egmontaz, Ejk81, Ejs-80, El C, El Suizo, El estremeñu, El xeneixe12, Electionworld, Elockid, Eloquence, Emet truth, Emil Kastberg, Emmisa, EmpireKing, Enchanter, EncycloPetey, Energyfreezer, EnglishDude, Enigmocracy, EoGuy, Epbr123, Epicadam, Equendil, Erianna, Eric-Wester, Ericd, Ericorides, Esrever, Esteffect, Estel, Eu.stefan, Euratlas, EurekaLott, Europeanul, Everyking, ExRat, Eztoremember, FF2010, FJM, FJPB, Faeyd, Falcon8765, Fan-1967, FatherofAssBurgler, Favonian, Fayenatic london, Fdorville, Felyza, Ferhengvan, Ferix, Ferkelparade, Filll, Flatterworld, FlavrSavr, Flewis, FlieGerFaUstMe262, Floul1, Flowerpotman, Fluri, Fonzy, Foreigner, Forest Elf, Forumyoungblood, Fotaun, FotoPhest, Fox, Fr34k, Fram, France3470, FrancoGG, Francs2000, Frasor, Fred Condo, Fredgoat, Freeminder, FreplySpang, Frichmon, Friginator, From-cary, Frukko, Fsotrain09, Fudoreaper, Fungusbrain, Funnyhat, Furrykef, Future Perfect at Sunrise, Fuzheado, Fæ, G.A.S, G00dfella613, G026r, GC, GCarty, GL, Gadfium, Gail, Gaius Octavius Princeps, Gaj777, Gary King, Gayjin, Gazpacho, Gdo01, Geni, Geoff.powers, Georgegeorgegeorge, Geraldo Perez, Gerrit, Gestumblindi, Gggh, GiZiBoNG, Giftlite, Gilliam, Gingerskyler, Gioto, Girolamo Savonarola, Glane23, Glaurung, Glenn, Glenndsymons, Gman124, GoetheFromm, Gogo Dodo, Golbez, Goldom, Golgofrinchian, Good Olfactory, Goodlucca, Graeme Bartlett, Grafen, Graham87, GrandSapin, GrandpaDoc, Granpuff, Grayshi, Greatestrowerever, Green Giant, Green caterpillar, Gregory Shantz, Grillo, Grimey109, Gringo300, Groovyfairyfloss, Grosscha, Grundig, Gryffindor, Gsarwa, Gscshoyru, Guanaco, Guliolopez, Gurch, Guy Peters, Guywithmako, Gwernol, Gwguffey, Gyrothepyro, Gzornenplatz, Gökhan, Gürbetaler, HAMM, HJ Mitchell, Habib87, Hadal, Hairy Dude, HalfShadow, Hanacy, Hannes Röst, Hans Dunkelberg, Harej, Harel Newman, Harp, HarrisMC, Harry the Dirty Dog, Harshapaulraj, Hateless, Hayden120, Hdstubbs, Headbomb, Hemnarayan, Henrygb, Hephaestos, HexaChord, Hgilbert, Hiddekel, Highvale, Hiimtod, Historian19, Hmains, Hmusseau, Hoary, Hooriaj, Horgen, Hotcrocodile, House.au, Howdydoo, Hu12, Huntington, Hushpuckena, Husond, Hut 8.5, Huwr, Hydrogen Iodide, Hyyttaa, IRP, Ianb, Ice Cold Beer, Icemuon, IchbinIch, Idaltu, Idinic, Ief, Iem, Igo4U, Immunize, Improv, InShaneee, Indon, Infrogmation, Inner Earth, Inside switzerland magazine, Insideswitzerland, Instinct, Intangible, Into The Fray, Ioannes Tzimiskes, Iridescent, IronGargoyle, IstvanWolf, Italiano111, ItsZippy, Itsmine, Iunaw, Ivirivi00, Ixfd64, Izzedine, J. 'mach' wust, J. Naven, J.Wright, J.delanoy, JForget, JFreeman, JNW, JVz, JYi, JaGa, Jab843, Jackalisaghost, Jackl, JackofOz, Jacobgibbons, Jacooks, Jaggsan, Jake Wartenberg, JakeVortex, Jakezing, Jakro64, Jam01, Jamcib, JamesR, JamesReyes, Jamiestahl, Jammann, Jao, Jaranda, Jauerback, Jayen466, Jayron32, Jbsongaku, Jcain1635, Jclemens, Jeanenawhitney, Jeff G., JeffDaniels, Jeremydillon, JerryFriedman, Jerryseinfeld, Jessejamesbisunia, Jezhotwells, Jhbuk, Jhendin, Jhjhjhffffffffffff, JiMidnite, Jiang, JimVC3, Jimspilman, Jj137, Jjhcap99, Jklumker, Jlmlax99, Jmaddux23, Joebob23, Joffeloff, John, John of Reading, Johnandbob, Johnnycakes234, Jojhutton, Jon cristaldi, JonMoore, Joolz, Jordz, Jorunn, Jose77, Joseph Solis in Australia, JoshHolloway, José Ignacio Cabrejos, Jovianeye, Joy, Joysofpi, Jpeob, Jpowell, Jrielaecher, Jsl83, Juliancolton, Jumbuck, Jørgen, K10wnsta, Kabelleger, Kaf, Kaihsu, Kain Nihil, Kairos, Kaiserb, KaitlinC11, Kaitsu80, Kanatonian, Kane5187, Kangy, Kannie, Karl, KarlFrei, Karuna8, Kat90210, Katalaveno, Kbh3rd, Kbolino, Keenan Pepper, Keilana, Keimzelle, Kelisi, Kelnos, Kenbei, Kessler, Kevin Murray, Keycard, Khoikhoi, Killervogel5, Kim-Zhang-Hong, King Toadsworth, King of Hearts, Kingjezza, Kingpin13, Kintetsubuffalo, Kiss My Ass Bitch, Kitch, Klapper, Kman665, KnowledgeOfSelf, Kokiri, Kokpit, Kosovarvirg, Kotniski, Koyaanis Qatsi, Kozuch, Kr1st1deejay97, Krator, Krellis, Kseferovic, Ksnow, Kubigula, Kusma, Kwaku, Kwamikagami, Kylu, Kyphe, KyraVixen, Kzeynel, L Kensington, L95slovenia, LAAFan, LAX, LFaraone, LOL, La Pianista, LaMenta3, Lairor, Landon1980, Latka, Lazyfair, Lazylaces, Leandrod, LeaveSleaves, LeedsKing, Leflyman, Legendsword, LeoDV, Leoni2, Leszek Jańczuk, Lethe, Levineps, Lezard, Lfgoette, Lfh, Lg1223, Liblamb, Liftarn, Lightmouse, Ligulem, Lijnema, Lilac Soul, Linburats, LineOfBlackCars, Lineface, Lithpiperpilot, Littrans, Lkaelin, Lkjhgfdsa, Lkmorlan, LlywelynII, Loadmaster, Lochness Monstah, Lollyz10, Longwayround, Lopokoko, Loren.wilton, LorenzoB, Lostcauses, Lostvalley, Lotje, Lprideux, Lsy098, Luca.gambazzi, Luffy487, Luna Santin, Lupin, Lupo, M0s6p, MBisanz, MER-C, MJCdetroit, MJD86, MKWat, MONGO, MPerel, MZMcBride, Mach2moo, MadGeographer, MadSurgeon, Madhero88, Maelnuneb, Magicalsaumy, Magore, Mahahahaneapneap, Male1979, MaliNorway, Mamawrites, Man vyi, Mandarax, Mangwanani, ManyFireflies, Maokart444, Marc Mongenet, Marco bisello,

Marcworca, Marek69, Markb, Markonen, Marktreut, Marlinite, Marlonparker, Mamanel, Martin451, MartinUK, Martinwilke1980, Marzoregulus@hotmail.com, Mascia Luccova, Masterjamie, Matdrodes, Mateo SA, Materialscientist, Mathpianist93, Matt Crypto, MattieTK, Maurog, Max rspct, MaxSem, Maziotis, Mbalelo, Mboverload, McDogm, McSly, Mclay1, Mcmuffinpie, Mcmuffinz, Mdw0, Meaghan, Media anthro, Meegs, Meekywiki, Meertn, Megaman3020, Melaen, Melsaran, Mentifisto, Mentisock, Meow kitties, Merbabu, Mericasio, Merovingian, Mesgul82, MetsFan76, Mezigue, Mic, Michael Hardy, MichaelBillington, Michelle1234567890987654321, Mickey gfss2007, Midright, Midwestern Demigod, Mifter, Mikaey, Mikaveikka, Mike12123, Mike15564, Mikeo, Mileage, Millahnna, MindstormsKid, Minhnd, Miquonranger03, Miraculousrandomness, Mirv, Mjgw, Mkilly, Mkztg, Mm.srb, Mmdoogie, Mo0, Momet, Momo erez, Monni1995, Montrealais, Monty845, Morana, Moreschi, Morio, Mossmen1531, Moumine, Mouse Nightshirt, Mr mormon, MrRadioGuy, Mrdeco, Mrholybrain, Mrxak, Msaini, Msikma, Muchosucko, Mufka, Mukitil, Mxcatania, Mxn, My76Strat, Mygerardromance, Myleslong, Mysdaao, Mzmujib, Mithrandir, N2e, N5iln, NByz, NCDane, NHRHS2010, NWill, Nach0king, Nagy, Naive cynic, Nakon, Nakos2208, Nathanael Bar-Aur L., Natl1, NawlinWiki, Nburden, Ncdoyle, Nellis, NeoChaosX, Nepenthes, NerdyNSK, Netkinetic, Netknowle, Neverquick, Nevilley, NewEnglandYankee, Ngagnebin, Niamh Smells, Nichalp, Nicholassharland, Nick, Nick Michael, Nickbenjohn1, Nightstallion, Nirvana2013, Nishant12, Nivix, NixonB, Njaelkies Lea, Njm0, Nkcs, Nlu, Nneonneo, Noah Salzman, Noisy, NonNobisSolum, Nonpawit, Nonplus, Nopetro, NorVegan, Nossac, Notredamefan747, Nova77, Nposs, Nsaa, NuclearWarfare, Nudibranch2, Nuggetman, Numbo3, OL144, OSborn, Oda Mari, Odengatan, Off2riorob, Ohconfucius, Ohnoitsjamie, Ohwell32, Olivier, Omicronpersei8, Oneilius, Ordew, Oreo Priest, Osterluzei, Ottawa4ever, Ourai, OwenBlacker, Ozes, P4k, PAK Man, PBS-AWB, PFHLai, PL290, PRhyu, PZFUN, Padraig, Paine Ellsworth, ParisianBlade, Parmaestro, Party, Pascar, Pasu, Pathoschild, PatriceNeff, Patrick, Patrickmayor, Patstuart, Pattycakes13, Paul August, Paul Faller, Paulyeobee88, Pavel Vozenilek, Pax:Vobiscum, PeaceNT, Pearle, Pecksique, Pensionero, Pepper, Peter Horn, Peter.C, PeterSymonds, Petersdick, Pethr, Petrb, Pg2114, Pgk, Pharaoh of the Wizards, Pharos04, PhilKnight, PhilMacD, Philip Trueman, Philippe317, Philthecow, Phoenix-forgotten, Pi, PiMaster3, Piano non troppo, Piccolo Modificatore Laborioso, Pigman, Piledhigheranddeeper, Pimplilboyahanicky, Pinethicket, Pippu d'Angelo, PizzaofDoom, Pizzarocks09, Pjetter, PlantHead, Pne, Pokymon99, Pol098, Polaron, Postmortemjapan, Postoak, Pozcircuitboy, Pp.zg.hr, PranksterTurtle, Prashanthns, Prince of History, PrincessofLlyr, Prodego, Proteus, Prylon, PsY.cHo, Psy guy, Ptah, the El Daoud, PurpleInk, Purpleturple, Pv2b, Pyrospirit, Qaka, Qasaqsuyu, Quadpus, Quercus basaseachicensis, Quiensabe, Quizimodo (usurped), Qwertasdf10, Qwyrxian, Qxz, R9tgokunks, RB972, RFBailey, RTC, Racerguy, RainbowOfLight, Raining girl, Rama, Ran, Random user 39849958, RandomAct, RandomOrca2, RandomStringOfCharacters, RaseaC, Ratemonth, Raven in Orbit, Raven4x4x, Ray Van De Walker, Razorflame, Rctycoplay, Rdsmith4, Rebug, Red Winged Duck, RedZebra, Redbombtoma, Redgolpe, Redhill54, Redmarkviolinist, Redsox11790, Redsxfenway, Regibox, Reinthal, Reisio, Rejectwater, Rejedef, Renaissancee, Renata, Revelian, Reverendgraham, Revolución, Revolutionary, RexNL, Rexlis, Reywas92, Riana, Rich Farmbrough, Rich257, Richard Arthur Norton (1958-), Richard New Forest, Rick Block, RickK, Ricky81682, Ricky@36, Rickyiskandar, RiingoNYM, Riker wt, Rizalninoynapoleon, Rjwilmsi, Rl, Rlbarton, Rmhermen, Robert25, Robert3, Robertissimo, Rocket000, Rockymountains, Rodney Boyd, Roland Abt, Roland Kaufmann, Romanm, Ron Ritzman, Ronhjones, Ronin of Chaos, Rotovia, RoyBoy, Rredwell, Rror, Rsrikanth05, Rudolfhess456, Ruhrjung, Rumpelstiltskin223, RussBlau, Ryan McGuinness, RyanGerbil10, S charette, SDC, SMC89, SSZ, ST47, Saforrest, Saimdusan, Saippuakauppias, Salaciousd, Salion, Salvio giuliano, Sam Blacketer, Samantha555, Sameer.charles, Samrolken, Samuel.Mago, Samwb123, Sandstein, Sanghvimihir, Sango123, Sannse, Satori Son, ScaldingHotSoup, Schneelocke, Schutz, Scipius, Sciurinæ, Scohoust, Scott Gall, Scwlong, Sdnegel, SeanLegassick, Seaphoto, Search4Lancer, Seb az86556, Seba5618, Sebesta, Secfan, Secretlondon, Seki rs, Selket, Sellyme, Semmler, Sengstag, Seric2, Serte, Sewing, Shadow007, Shadowblade, Shadowslayer978, Shak-74M, Shamir1, Shandris, Sharkface217, Shawn in Montreal, Shawn2082, Shlomke, Shoessss, Shortstraw, Shtop Shtealing from my Shtore, Sidonius, SigPig, Sikaolli, Silly rabbit, Silvermoonx, Silverxxx, Simcom, Simkiott, SimonP, Sims2aholic8, Sin-man, SineWave, Singularity, Sinn, Siqbal, Sir Richardson, Sjakkalle, Sk8er777, Skew-t, Skizzik, Skunkboy74, Skunkwaffle, SkyBlue eagle, Slowmover, SluggoOne, Sluzzelin, Slysplace, Smackdownvsraw, SmellyLilYou, Smirnius, SnowFire, Snowolf, Soelke, Soman, SomeHuman, Somnewise@yahoo.com, Soumyasch, Sourside21, South Bay, Southparkfan007, SpNeo, Sparr, Spellcast, Spettro9, SpookyMulder, Springnuts, Sputniktilt, Srrostum, Srushe, Staffwaterboy, Stalker X, Stan Shebs, Startstop123, Station1, Steel, Steinsky, Stemonitis, Stephan Leeds, Stephan Schulz, Stephenb, Stepheng3, Stevertigo, StradivariusTV, Strafpeloton2, Stuinzuri, Stupid Corn, Stw, Sue Rangell, Suitequai, Sullivanfam, Summer Song, Sunray, Supercoop, Supertask, Sureshot54, Suruena, Suse, Sven Manguard, Svenman, Svizra, Swapper, Swi521, Swid, Swiss Ace, SwissCelt, SwissTHX11384EB, Switzerlandinsound, Swordfish2008, Synchronism, Synthetik, Systemas, Szlam, Szwi, T.M.M. Dowd, TERIKATA, TFOWR, TGC55, THF, TLRkahuna, TUF-KAT, Ta bu shi da yu, Tad Lincoln, Taemyr, Tang Wenlong, Tangotango, Tanthalas39, Targeman, Tavis, Tbhotch, Tcc9mpl, Tedickey, TedlyW, Teemu Ruskeepää, Teerickson, Teiladnam, Tenmei, Tennis expert, Tequendamia, TerreOcre, TestPilot, Tetracube, Thadius856, Thatdog, The Egyptian Liberal, The Epopt, The Evil Spartan, The High Fin Sperm Whale, The Thing That Should Not Be, The Wild West guy, The Wordsmith, The sanch, The undertow, TheCatalyst31, TheKMan, TheProject, Theanthrope, Thecurran, Thedarxide, Thedjatclubrock, Thejason, Thelb4, Themanwithoutapast, Therequiembellishere, Thingg, Think outside the box, ThomasK, Thomasankers, Thricecube, Thue, Tiddly Tom, Tide rolls, TiffaF, Tiggerjay, Tillwe, Tim bates, TimBentley, Timir2, Timlong28, Timrollpickering, Timwi, Tiptoety, Tirerim, Titoxd, Tjayh913, Tnxman307, Tobby72, Tobias Galtieri, Tobyc75, Todowd, Tofubar, Tohd8BohaithuGh1, Tom Radulovich, TomH, TomPreuss, Tombomp, Tomchen1989, Tommy2010, Tony Fox, Tony1, Tonyle, Tosão, TouristPhilosopher, Toussaint, Tox, Tpbradbury, TreasuryTag, Trigaranus, Trilobite, Triona, Trish Boland, TriviaKing, TrixKidRox, Trompeta, Trusilver, Tuomas, Turkuun, Twsx, Tye, UKe-CH, Jirauna, UkPaolo, Ukryan1992, Ulmke, Ulric1313, Ulritz, Uncle Milty, Uncle Scrooge, UncleDouggie, Unyoyega, User179514, Uusef, VI, VX-LTM, Vacekha, Van helsing, Vanished 6551232, Vardion, Vasilcho, Vcelloho, Veledan, Veritologist, Vettrock, Vianello, Victor falk, Victorjwilli, Vinni3, Vivisfugue, Vnsdvjnsvksnvjkmc, VolatileChemical, Vortexrealm, Voyager77377, Voyagerfan5761, Vrenator, Vsmith, Vzbs34, WTGDMan1986, WadeSimMiser, Wafulz, Waggers, Wakablogger2, Waldir, Walrus heart, Wangi, WarthogDemon, Washburnmav, Wayne Slam, Wbennettjr, Wbkelley, Welsh, Weregerbil, Wereon, WhisperToMe, Whoda, Whosyourjudas, Wik, WikHead, Wiki Raja, Wiki alf, Wiki-vr, Wiki.Tango.Foxtrot, WikiLaurent, Wikialthaf, Wikibob, Wikid77, Wikidemon, Wikidenizen, Wikieditor06, Wikiepdiax818, Wikievil666, Wikipelli, Wikiscribe, Wikiturk, William Avery, Willking1979, Wimt, Wisdom89, Wmahan, WolfgangFaber, Wombat81, Woohookitty, World, Wormcast, Wprlh, Writtenright, X0man0x, X2wolf, Xaosflux, Xezbeth, Xonqnopp, Xyzzy n, Xyzzyva, Yaf, Yahia.barie, Yamamoto Ichiro, Yardbirdjo, Ybruf, YellowMonkey, Ynilp, Yorkshirian, Yvesnimmo, Ywmpq205, Z.E.R.O., Zabadinho, Zacharie Grossen, Zekerags, Zenosparadox, ZeppelinFan29, Zidonuke, Zntrip, Zoe, Zondor, ZooFari, Zscout370, Zurichtourismus, Zzuuzz, ÇŸ³â°è±◆, Île flottante, Александър, Гаджиев Азер, , 3612 anonymous edits

Aar *Source*: http://en.wikipedia.org/w/index.php?title=Aar *Contributors*: 24ip, 3122WIKI, Ahoerstemeier, Ami in CH, Andre Engels, Andres, Antti29, Ashot Gabrielyan, Attilios, BD2412, Badener, BillC, Bob Burkhardt, Cantiorix, Ceyockey, Claude girardin, Conversion script, D6, Darwinek, Dbachmann, Dcooper, Docu, Dr. Blofeld, Emperor, Eob, Epbr123, Fanyavizuri, Flying Saucer, Gaius Cornelius, Gene Nygaard, GlobeGores, Gwern, Harleyrules170, Hottentot, Inwind, Irmgard, Isnow, Japanese Searobin, Jerzy, Jyril, Kokiri, Lee Daniel Crocker, LeoNomis, Leonard Vertighel, Lightmouse, Lincher, MadGeographer, Magnum1, Mark Wheaver, Mav, Maximaximax, MightyWarrior, Mimihitam, Mmounties, Mrmoocow, NickCT, Olessi, Olivier, Ori, P-A., Paracel63, Paranoid, Pedant17, PhilKnight, Photnart, Piotrus, Pmsyyz, Razdan1974, Rjwilmsi, Robert Horning, S, ST47, SailorfromNH, Sandstein, Skapur, Sligocki, Sluzzelin, Squids and Chips, Tobyc75, TracksOnWax, Ulamm, Verne Equinox, Veyklevar, Wetman, WhisperToMe, Wiikid, Wikibob, Zacharie Grossen, Zumbo, 22 anonymous edits

Image Sources, Licenses and Contributors

file:Klingnauer Stausee 2008 10 17 000.JPG *Source*: http://en.wikipedia.org/w/index.php?title=File:Klingnauer_Stausee_2008_10_17_000.JPG *License*: unknown *Contributors*: User:MichaelFrey, User:Sundance Raphael

File:Schloss Böttstein Innenhof.JPG *Source*: http://en.wikipedia.org/w/index.php?title=File:Schloss_Böttstein_Innenhof.JPG *License*: unknown *Contributors*: User:MichaelFrey

File:Coat of arms of Boettstein.svg *Source*: http://en.wikipedia.org/w/index.php?title=File:Coat_of_arms_of_Boettstein.svg *License*: unknown *Contributors*: diese Datei: Jwnabd

file:Switzerland location map.svg *Source*: http://en.wikipedia.org/w/index.php?title=File:Switzerland_location_map.svg *License*: unknown *Contributors*: Lukasb1992, NordNordWest, Sting, 1 anonymous edits

File:Red pog.svg *Source*: http://en.wikipedia.org/w/index.php?title=File:Red_pog.svg *License*: unknown *Contributors*: Anomie

Image:Population Boettstein.svg *Source*: http://en.wikipedia.org/w/index.php?title=File:Population_Boettstein.svg *License*: unknown *Contributors*: User:MichaelFrey

Image:Schloss Böttstein Tor 1.JPG *Source*: http://en.wikipedia.org/w/index.php?title=File:Schloss_Böttstein_Tor_1.JPG *License*: unknown *Contributors*: User:MichaelFrey

File:Karte Bezirk Zurzach 2007.png *Source* http://en.wikipedia.org/w/index.php?title=File:Karte_Bezirk_Zurzach_2007.png *License*: unknown *Contributors*: Tschubby

File:Flag of Switzerland.svg *Source*: http://en.wikipedia.org/w/index.php?title=File:Flag_of_Switzerland.svg *License*: unknown *Contributors*: User:-xfi-, User:Marc Mongenet, User:Zscout370

File:Flag of Canton of Aargau.svg *Source*: http://en.wikipedia.org/w/index.php?title=File:Flag_of_Canton_of_Aargau.svg *License*: unknown *Contributors*: Anaty, Anime Addict AA, Damouns, Denelson83, DenghiùComm, Fry1989, Mattes, ZorkNika, 1 anonymous edits

Image:Baldingen-blason.png *Source*: http://en.wikipedia.org/w/index.php?title=File:Baldingen-blason.png *License*: unknown *Contributors*: User:Voyager

Image:Boebikon-blason.png *Source*: http://en.wikipedia.org/w/index.php?title=File:Boebikon-blason.png *License*: unknown *Contributors*: User:Voyager

Image:Boettstein-blason.png *Source*: http://en.wikipedia.org/w/index.php?title=File:Boettstein-blason.png *License*: unknown *Contributors*: User:Voyager

Image:Doettingen-blason.png *Source*: http://en.wikipedia.org/w/index.php?title=File:Doettingen-blason.png *License*: unknown *Contributors*: User:Voyager

Image:Endingen-blason.png *Source*: http://en.wikipedia.org/w/index.php?title=File:Endingen-blason.png *License*: unknown *Contributors*: User:Voyager

Image:Fisibach-blason.png *Source*: http://en.wikipedia.org/w/index.php?title=File:Fisibach-blason.png *License*: unknown *Contributors*: User:Voyager

Image:Full-Reuenthal-blason.png *Source*: http://en.wikipedia.org/w/index.php?title=File:Full-Reuenthal-blason.png *License*: unknown *Contributors*: User:Voyager

Image:Kaiserstuhl-blason.png *Source*: http://en.wikipedia.org/w/index.php?title=File:Kaiserstuhl-blason.png *License*: unknown *Contributors*: User:Voyager

Image:Klingnau-blason.png *Source*: http://en.wikipedia.org/w/index.php?title=File:Klingnau-blason.png *License*: unknown *Contributors*: User:Voyager

Image:Koblenz-blason.png *Source*: http://en.wikipedia.org/w/index.php?title=File:Koblenz-blason.png *License*: unknown *Contributors*: User:Voyager

Image:Leibstadt-blason.png *Source*: http://en.wikipedia.org/w/index.php?title=File:Leibstadt-blason.png *License*: unknown *Contributors*: User:Voyager

Image:Lengnau-blason.png *Source*: http://en.wikipedia.org/w/index.php?title=File:Lengnau-blason.png *License*: unknown *Contributors*: User:Voyager

Image:Leuggern-blason.png *Source*: http://en.wikipedia.org/w/index.php?title=File:Leuggern-blason.png *License*: unknown *Contributors*: User:Voyager

Image:Mellikon-blason.png *Source*: http://en.wikipedia.org/w/index.php?title=File:Mellikon-blason.png *License*: unknown *Contributors*: User:Voyager

Image:Rekingen-blason.png *Source*: http://en.wikipedia.org/w/index.php?title=File:Rekingen-blason.png *License*: unknown *Contributors*: User:Voyager

Image:Rietheim-blason.png *Source*: http://en.wikipedia.org/w/index.php?title=File:Rietheim-blason.png *License*: unknown *Contributors*: User:Voyager

Image:Ruemikon-blason.png *Source*: http://en.wikipedia.org/w/index.php?title=File:Ruemikon-blason.png *License*: unknown *Contributors*: User:Voyager

Image:Schneisingen-blason.png *Source*: http://en.wikipedia.org/w/index.php?title=File:Schneisingen-blason.png *License*: unknown *Contributors*: User:Voyager

Image:Siglistorf-blason.png *Source*: http://en.wikipedia.org/w/index.php?title=File:Siglistorf-blason.png *License*: unknown *Contributors*: User:Voyager

Image:Tegerfelden-blason.png *Source*: http://en.wikipedia.org/w/index.php?title=File:Tegerfelden-blason.png *License*: unknown *Contributors*: User:Voyager

Image:Unterendingen-blason.png *Source*: http://en.wikipedia.org/w/index.php?title=File:Unterendingen-blason.png *License*: unknown *Contributors*: User:Voyager

Image:Wislikofen-blason.png *Source*: http://en.wikipedia.org/w/index.php?title=File:Wislikofen-blason.png *License*: unknown *Contributors*: User:Voyager

File:Switzerland-Baden-oldcityview.jpg *Source*: http://en.wikipedia.org/w/index.php?title=File:Switzerland-Baden-oldcityview.jpg *License*: unknown *Contributors*: User:Monk

File:Coat of arms of Baden AG.svg *Source*: http://en.wikipedia.org/w/index.php?title=File:Coat_of_arms_of_Baden_AG.svg *License*: unknown *Contributors*: diese Datei: Jwnabd

File:Wettingen Isistempel.jpg *Source*: http://en.wikipedia.org/w/index.php?title=File:Wettingen_Isistempel.jpg *License*: unknown *Contributors*: User:Badener

File:Merian Baden AG 1642.jpg *Source*: http://en.wikipedia.org/w/index.php?title=File:Merian_Baden_AG_1642.jpg *License*: unknown *Contributors*: User:Sidonius

File:Baden Altstadt 9502.JPG *Source*: http://en.wikipedia.org/w/index.php?title=File:Baden_Altstadt_9502.JPG *License*: unknown *Contributors*: User:Badener

File:Baden bahnhof.jpg *Source*: http://en.wikipedia.org/w/index.php?title=File:Baden_bahnhof.jpg *License*: unknown *Contributors*: Badener, Voyager

File:Baden Stein 9667.jpg *Source*: http://en.wikipedia.org/w/index.php?title=File:Baden_Stein_9667.jpg *License*: unknown *Contributors*: User:Badener

File:Stadtkirche Baden AG 7870.jpg *Source*: http://en.wikipedia.org/w/index.php?title=File:Stadtkirche_Baden_AG_7870.jpg *License*: unknown *Contributors*: User:Badener

File:Baden AG Ref Kirche.jpg *Source*: http://en.wikipedia.org/w/index.php?title=File:Baden_AG_Ref_Kirche.jpg *License*: unknown *Contributors*: Lutz Fischer-Lamprecht

File:Langmatt1.jpg *Source*: http://en.wikipedia.org/w/index.php?title=File:Langmatt1.jpg *License*: unknown *Contributors*: User:Irmgard

File:Baden Stadthaus.jpg *Source*: http://en.wikipedia.org/w/index.php?title=File:Baden_Stadthaus.jpg *License*: unknown *Contributors*: User:Voyager

File:Tagsatzung1531.jpg *Source*: http://en.wikipedia.org/w/index.php?title=File:Tagsatzung1531.jpg *License*: unknown *Contributors*: Fransvannes, Man vyi, Pymouss, Roland zh, Ronaldino, Voyager

File:Baden landvogtei.jpg *Source*: http://en.wikipedia.org/w/index.php?title=File:Baden_landvogtei.jpg *License*: unknown *Contributors*: Irmgard, Kurpfalzbilder.de, Manoillon, Ronaldino, Voyager, 1 anonymous edits

File:Baden KraftwerkAue.jpg *Source*: http://en.wikipedia.org/w/index.php?title=File:Baden_KraftwerkAue.jpg *License*: unknown *Contributors*: User:Voyager

File:Baden Aufnahmegebaude.jpg *Source*: http://en.wikipedia.org/w/index.php?title=File:Baden_Aufnahmegebaude.jpg *License*: unknown *Contributors*: User:Ikiwaner

File:Hallstatt LaTene.png *Source*: http://en.wikipedia.org/w/index.php?title=File:Hallstatt_LaTene.png *License*: unknown *Contributors*: User:Dbachmann

File:Bund-ro-altburg.jpg *Source*: http://en.wikipedia.org/w/index.php?title=File:Bund-ro-altburg.jpg *License*: unknown *Contributors*: Original uploader was Roßbacher at de.wikipedia

File:Romano-Celtic mirror (Desborough).jpg *Source*: http://en.wikipedia.org/w/index.php?title=File:Romano-Celtic_mirror_(Desborough).jpg *License*: unknown *Contributors*: User:Fuzzypeg

File:Hallstatt.png *Source*: http://en.wikipedia.org/w/index.php?title=File:Hallstatt.png *License*: unknown *Contributors*: User:Hxseek

File:Cratère de Vix 0023.jpg *Source*: http://en.wikipedia.org/w/index.php?title=File:Cratère_de_Vix_0023.jpg *License*: unknown *Contributors*: Bohème, Esprit Fugace, Johnbod, Manchot, Marcus Cyron, Siren-Com, Zolo

File:Wappen Aargau matt.svg *Source*: http://en.wikipedia.org/w/index.php?title=File:Wappen_Aargau_matt.svg *License*: unknown *Contributors*: Sa-se

File:Karte Lage Kanton Aargau 2011 2.png *Source*: http://en.wikipedia.org/w/index.php?title=File:Karte_Lage_Kanton_Aargau_2011_2.png *License*: unknown *Contributors*: Tschubby

File:Aarau Altstadt.jpg *Source*: http://en.wikipedia.org/w/index.php?title=File:Aarau_Altstadt.jpg *License*: unknown *Contributors*: Breeze, Docu, EDUCA33E, Slomox, Voyager, 1 anonymous edits

File:Bözberg Lägernblick 8054.jpg *Source*: http://en.wikipedia.org/w/index.php?title=File:Bözberg_Lägernblick_8054.jpg *License*: unknown *Contributors*: Benutzer:BadenerBadener

File:Karte Kanton Aargau Bezirke.png *Source*: http://en.wikipedia.org/w/index.php?title=File:Karte_Kanton_Aargau_Bezirke.png *License*: unknown *Contributors*: Juiced lemon, Tschubby

File:KKL Leibstadt.jpg *Source*: http://en.wikipedia.org/w/index.php?title=File:KKL_Leibstadt.jpg *License*: unknown *Contributors*: User:Nawi112

File:Flag of Switzerland (Pantone).svg *Source*: http://en.wikipedia.org/w/index.php?title=File:Flag_of_Switzerland_(Pantone).svg *License*: unknown *Contributors*: User:-xfi-, User:Marc Mongenet, User:Zscout370

File:Coat of Arms of Switzerland (Pantone).svg *Source*: http://en.wikipedia.org/w/index.php?title=File:Coat_of_Arms_of_Switzerland_(Pantone).svg *License*: unknown *Contributors*: User:E Pluribus Anthony, User:Reisio, User:Wiki-vr, User:Zscout370

File:Europe-Switzerland.svg *Source*: http://en.wikipedia.org/w/index.php?title=File:Europe-Switzerland.svg *License*: unknown *Contributors*: User:Hayden120, User:NuclearVacuum
File:Increase2.svg *Source*: http://en.wikipedia.org/w/index.php?title=File:Increase2.svg *License*: unknown *Contributors*: Sarang
File:Theater Kaiseraugst.jpg *Source*: http://en.wikipedia.org/w/index.php?title=File:Theater_Kaiseraugst.jpg *License*: unknown *Contributors*: User:Taxiarchos228
File:Bundesbrief.jpg *Source*: http://en.wikipedia.org/w/index.php?title=File:Bundesbrief.jpg *License*: unknown *Contributors*: AnRo0002, AndreasPraefcke, Ekki01, Lupo, Parpan05, Paul2, Sidonius
File:Old Swiss Confederation.jpg *Source*: http://en.wikipedia.org/w/index.php?title=File:Old_Swiss_Confederation.jpg *License*: unknown *Contributors*: User:Lupo
File:Acte de Médiation mg 0643.jpg *Source*: http://en.wikipedia.org/w/index.php?title=File:Acte_de_Médiation_mg_0643.jpg *License*: unknown *Contributors*: User:Rama
File:Bern, Federal Palace, 1857.jpg *Source*: http://en.wikipedia.org/w/index.php?title=File:Bern,_Federal_Palace,_1857.jpg *License*: unknown *Contributors*: Ch. Durheim
File:Gotthard Eröffnungszug Bellinzona.jpg *Source*: http://en.wikipedia.org/w/index.php?title=File:Gotthard_Eröffnungszug_Bellinzona.jpg *License*: unknown *Contributors*: User:Parpan05
File:Ulrich Wille.jpg *Source*: http://en.wikipedia.org/w/index.php?title=File:Ulrich_Wille.jpg *License*: unknown *Contributors*: Dbachmann, Kresspahl, Roland zh
File:Bundesrat Conseil Consglio Cussegl.jpg *Source*: http://en.wikipedia.org/w/index.php?title=File:Bundesrat_Conseil_Consglio_Cussegl.jpg *License*: unknown *Contributors*: Swiss Federal Council
File:Satellite image of Switzerland in September 2002.jpg *Source*: http://en.wikipedia.org/w/index.php?title=File:Satellite_image_of_Switzerland_in_September_2002.jpg *License*: unknown *Contributors*: Hautala, Tschubby, Zacharie Grossen
Image:Hauteroute.jpg *Source*: http://en.wikipedia.org/w/index.php?title=File:Hauteroute.jpg *License*: unknown *Contributors*: User:Jackph
Image:Sanetsch.jpg *Source*: http://en.wikipedia.org/w/index.php?title=File:Sanetsch.jpg *License*: unknown *Contributors*: User:Terra3
Image:Central Switzerland.jpg *Source*: http://en.wikipedia.org/w/index.php?title=File:Central_Switzerland.jpg *License*: unknown *Contributors*: User:Clare66
Image:Aletschhorn from Konkordiaplatz.jpg *Source*: http://en.wikipedia.org/w/index.php?title=File:Aletschhorn_from_Konkordiaplatz.jpg *License*: unknown *Contributors*: User:Tallin
Image:Joux.jpg *Source*: http://en.wikipedia.org/w/index.php?title=File:Joux.jpg *License*: unknown *Contributors*: User:Earth explorer
Image:Lago di Lugano3.jpg *Source*: http://en.wikipedia.org/w/index.php?title=File:Lago_di_Lugano3.jpg *License*: unknown *Contributors*: User:MadGeographer
File:Bundesrat der Schweiz 2011-H25P.jpg *Source*: http://en.wikipedia.org/w/index.php?title=File:Bundesrat_der_Schweiz_2011-H25P.jpg *License*: unknown *Contributors*: Swiss Federal Council Photographer: Monika Flückiger
File:Bern 104.jpg *Source*: http://en.wikipedia.org/w/index.php?title=File:Bern_104.jpg *License*: unknown *Contributors*: Pazit Polak
File:Landsgemeinde Glarus 2006.jpg *Source*: http://en.wikipedia.org/w/index.php?title=File:Landsgemeinde_Glarus_2006.jpg *License*: unknown *Contributors*: Adrian Sulc
File:Suisse cantons.svg *Source*: http://en.wikipedia.org/w/index.php?title=File:Suisse_cantons.svg *License*: unknown *Contributors*: User:Pymouss44
File:Wappen Nidwalden matt.svg *Source*: http://en.wikipedia.org/w/index.php?title=File:Wappen_Nidwalden_matt.svg *License*: unknown *Contributors*: Sa-se
File:Wappen Appenzell Ausserrhoden matt.svg *Source*: http://en.wikipedia.org/w/index.php?title=File:Wappen_Appenzell_Ausserrhoden_matt.svg *License*: unknown *Contributors*: Sa-se
File:Wappen Obwalden matt.svg *Source*: http://en.wikipedia.org/w/index.php?title=File:Wappen_Obwalden_matt.svg *License*: unknown *Contributors*: Sa-se
File:Wappen Appenzell Innerrhoden matt.svg *Source*: http://en.wikipedia.org/w/index.php?title=File:Wappen_Appenzell_Innerrhoden_matt.svg *License*: unknown *Contributors*: Sa-se
File:Wappen Schaffhausen matt.svg *Source*: http://en.wikipedia.org/w/index.php?title=File:Wappen_Schaffhausen_matt.svg *License*: unknown *Contributors*: Sa-se
File:Wappen Basel-Landschaft matt.svg *Source*: http://en.wikipedia.org/w/index.php?title=File:Wappen_Basel-Landschaft_matt.svg *License*: unknown *Contributors*: User:WIKImaniac
File:Wappen Schwyz matt.svg *Source*: http://en.wikipedia.org/w/index.php?title=File:Wappen_Schwyz_matt.svg *License*: unknown *Contributors*: -
File:Wappen Basel-Stadt matt.svg *Source*: http://en.wikipedia.org/w/index.php?title=File:Wappen_Basel-Stadt_matt.svg *License*: unknown *Contributors*: Sa-se
File:Wappen Solothurn matt.svg *Source*: http://en.wikipedia.org/w/index.php?title=File:Wappen_Solothurn_matt.svg *License*: unknown *Contributors*: Sa-se
File:Wappen Bern matt.svg *Source*: http://en.wikipedia.org/w/index.php?title=File:Wappen_Bern_matt.svg *License*: unknown *Contributors*: User:Sa-se
File:Wappen St. Gallen matt.svg *Source*: http://en.wikipedia.org/w/index.php?title=File:Wappen_St._Gallen_matt.svg *License*: unknown *Contributors*: Sa-se
File:Wappen Freiburg matt.svg *Source*: http://en.wikipedia.org/w/index.php?title=File:Wappen_Freiburg_matt.svg *License*: unknown *Contributors*: Sa-se
File:Wappen Thurgau matt.svg *Source*: http://en.wikipedia.org/w/index.php?title=File:Wappen_Thurgau_matt.svg *License*: unknown *Contributors*: Sa-se
File:Wappen Genf matt.svg *Source*: http://en.wikipedia.org/w/index.php?title=File:Wappen_Genf_matt.svg *License*: unknown *Contributors*: Sa-se
File:Wappen Tessin matt.svg *Source*: http://en.wikipedia.org/w/index.php?title=File:Wappen_Tessin_matt.svg *License*: unknown *Contributors*: Sa-se
File:Wappen Glarus matt.svg *Source*: http://en.wikipedia.org/w/index.php?title=File:Wappen_Glarus_matt.svg *License*: unknown *Contributors*: Sa-se
File:Wappen Uri matt.svg *Source*: http://en.wikipedia.org/w/index.php?title=File:Wappen_Uri_matt.svg *License*: unknown *Contributors*: Sa-se
File:Wappen Graubünden matt.svg *Source*: http://en.wikipedia.org/w/index.php?title=File:Wappen_Graubünden_matt.svg *License*: unknown *Contributors*: Sa-se
File:Wappen Wallis matt.svg *Source*: http://en.wikipedia.org/w/index.php?title=File:Wappen_Wallis_matt.svg *License*: unknown *Contributors*: Sa-se
File:Wappen Jura matt.svg *Source*: http://en.wikipedia.org/w/index.php?title=File:Wappen_Jura_matt.svg *License*: unknown *Contributors*: Sa-se
File:Wappen Waadt matt.svg *Source*: http://en.wikipedia.org/w/index.php?title=File:Wappen_Waadt_matt.svg *License*: unknown *Contributors*: Sa-se
File:Wappen Luzern matt.svg *Source*: http://en.wikipedia.org/w/index.php?title=File:Wappen_Luzern_matt.svg *License*: unknown *Contributors*: Sa-se
File:Wappen Zug matt.svg *Source*: http://en.wikipedia.org/w/index.php?title=File:Wappen_Zug_matt.svg *License*: unknown *Contributors*: Sa-se
File:Wappen Neuenburg matt.svg *Source*: http://en.wikipedia.org/w/index.php?title=File:Wappen_Neuenburg_matt.svg *License*: unknown *Contributors*: Sa-se
File:Wappen Zürich matt.svg *Source*: http://en.wikipedia.org/w/index.php?title=File:Wappen_Zürich_matt.svg *License*: unknown *Contributors*: Sa-se
File:Geneve2.jpg *Source*: http://en.wikipedia.org/w/index.php?title=File:Geneve2.jpg *License*: unknown *Contributors*: User:MadGeographer
File:Flag of the Red Cross.svg *Source*: http://en.wikipedia.org/w/index.php?title=File:Flag_of_the_Red_Cross.svg *License*: unknown *Contributors*: User:Jhs
File:Axalp FA-18C 5.jpg *Source*: http://en.wikipedia.org/w/index.php?title=File:Axalp_FA-18C_5.jpg *License*: unknown *Contributors*: User:Cooper.ch
File:SKdt-Fahrzeug - Schweizer Armee - Steel Parade 2006.jpg *Source*: http://en.wikipedia.org/w/index.php?title=File:SKdt-Fahrzeug_-_Schweizer_Armee_-_Steel_Parade_2006.jpg *License*: unknown *Contributors*: User:Sandstein
File:Vintage Omega Speedmaster 145.012-67.jpg *Source*: http://en.wikipedia.org/w/index.php?title=File:Vintage_Omega_Speedmaster_145.012-67.jpg *License*: unknown *Contributors*: Shane Lin
File:2011-06-14 08-28-11 Switzerland Kanton Zürich Zürich Hönggerberg.jpg *Source*: http://en.wikipedia.org/w/index.php?title=File:2011-06-14_08-28-11_Switzerland_Kanton_Zürich_Zürich_Hönggerberg.jpg *License*: unknown *Contributors*: User:Simisa
File:Engadine.jpg *Source*: http://en.wikipedia.org/w/index.php?title=File:Engadine.jpg *License*: unknown *Contributors*: User:Biovit
File:Swiss scientists.jpg *Source*: http://en.wikipedia.org/w/index.php?title=File:Swiss_scientists.jpg *License*: unknown *Contributors*: User:MadGeographer
File:ETHZ.JPG *Source*: http://en.wikipedia.org/w/index.php?title=File:ETHZ.JPG *License*: unknown *Contributors*: User:ZachT
File:LHC, CERN.jpg *Source*: http://en.wikipedia.org/w/index.php?title=File:LHC,_CERN.jpg *License*: unknown *Contributors*: Max Braun on Flickr
File:KKG-SW-Flight.jpg *Source*: http://en.wikipedia.org/w/index.php?title=File:KKG-SW-Flight.jpg *License*: unknown *Contributors*: Aare-Tessin AG für Elektrizität
File:Lötschberg Tunnel.jpg *Source*: http://en.wikipedia.org/w/index.php?title=File:Lötschberg_Tunnel.jpg *License*: unknown *Contributors*: User:Ttrainer
File:Sprachen CH 2000 EN.svg *Source*: http://en.wikipedia.org/w/index.php?title=File:Sprachen_CH_2000_EN.svg *License*: unknown *Contributors*: User:Sidonius
File:Bisse de Clavau.jpg *Source*: http://en.wikipedia.org/w/index.php?title=File:Bisse_de_Clavau.jpg *License*: unknown *Contributors*: Londo Mollari
File:Glarus Stadtkirche Langhaus.jpg *Source*: http://en.wikipedia.org/w/index.php?title=File:Glarus_Stadtkirche_Langhaus.jpg *License*: unknown *Contributors*: User:Ikiwaner
File:Vals06.JPG *Source*: http://en.wikipedia.org/w/index.php?title=File:Vals06.JPG *License*: unknown *Contributors*: AndreasPraefcke, Manoillon, Paebi, Wst, 1 anonymous edits
File:Rousseau Geneve.JPG *Source*: http://en.wikipedia.org/w/index.php?title=File:Rousseau_Geneve.JPG *License*: unknown *Contributors*: myself
File:Allalinhorn 4027 2007 04 17.JPG *Source*: http://en.wikipedia.org/w/index.php?title=File:Allalinhorn_4027_2007_04_17.JPG *License*: unknown *Contributors*: ch-info.ch
File:Innenaufnahme Vaillant Arena Davos.JPG *Source*: http://en.wikipedia.org/w/index.php?title=File:Innenaufnahme_Vaillant_Arena_Davos.JPG *License*: unknown *Contributors*: Original uploader was Abercio at de.wikipedia (Original text : Michael Petri)

File:Federer Cincinnati (2007).jpg *Source*: http://en.wikipedia.org/w/index.php?title=File:Federer_Cincinnati_(2007).jpg *License*: unknown *Contributors*: jtbarrett

File:CH Bern Aare pano-4.jpg *Source*: http://en.wikipedia.org/w/index.php?title=File:CH_Bern_Aare_pano-4.jpg *License*: unknown *Contributors*: User:Dschwen

File:Aar bassin.svg *Source*: http://en.wikipedia.org/w/index.php?title=File:Aar_bassin.svg *License*: unknown *Contributors*: User:Pymouss44

File:Aletschgebiet aus dem Flugzeug.jpg *Source*: http://en.wikipedia.org/w/index.php?title=File:Aletschgebiet_aus_dem_Flugzeug.jpg *License*: unknown *Contributors*: User:Robiand

File:Ausgang Aareschlucht.jpg *Source*: http://en.wikipedia.org/w/index.php?title=File:Ausgang_Aareschlucht.jpg *License*: unknown *Contributors*: Falk Lademann

File:Aareschlucht 166 7.jpg *Source*: http://en.wikipedia.org/w/index.php?title=File:Aareschlucht_166_7.jpg *License*: unknown *Contributors*: Pazit Polak

File:Aare - Limmat (Limmatspitz) IMG 6765.jpg *Source*: http://en.wikipedia.org/w/index.php?title=File:Aare_-_Limmat_(Limmatspitz)_IMG_6765.jpg *License*: unknown *Contributors*: User:Roland zh

File:London2007zurich img 5743.jpg *Source*: http://en.wikipedia.org/w/index.php?title=File:London2007zurich_img_5743.jpg *License*: unknown *Contributors*: User:Gürkan Sengün